SAXON MATH™

Course 1

Stephen Hake

Power-Up Workbook

www.SaxonPublishers.com
1-800-284-7019

ISBN 13: 978-1-5914-1823-8

Printed in the United States of America

8 0928 14 13 12 11 10
4500268053

Dear Student,

We enjoy watching the adventures of "Super Heroes" because they have power and they use their powers for good. Power is the ability to get things done. We acquire power through concentrated effort and practice. We build powerful bodies with vigorous exercise and healthy living. We develop powerful minds by learning and using skills that help us understand the world around us and solve problems that come our way.

We can build our mathematical power several ways. We can use our memory to store and instantly recall frequently used information. We can improve our ability to solve many kinds of problems mentally without using pencil and paper or a calculator. We can also expand the range of strategies we use to approach and solve new problems.

The Power Up section of each lesson in *Saxon Math Course 1* is designed to build your mathematical power. Each Power Up has three parts, Facts Practice, Mental Math, and Problem Solving. The three parts are printed on every Power Up page where you will record your answers. This workbook contains a Power Up page for every lesson.

Facts Practice is like a race—write the answers as fast as you can without making mistakes. If the information in the Fact Practice is new to you, take time to study the information so that you can recall the facts quickly and can complete the exercise faster next time.

Mental Math is the ability to work with numbers in your head. This skill greatly improves with practice. Each lesson includes several mental math problems. Your teacher will read these to you or ask you to read them in your book. Do your best to find the answer to each problem without using pencil and paper, except to record your answers. Strong mental math ability will help you throughout your life.

Problem Solving is like a puzzle. You need to figure out how to solve the puzzle. There are many different strategies you can use to solve problems. There are also some questions you can ask yourself to better understand a problem and come up with a plan to solve it. Your teacher will guide you through the problem each day. Becoming a good problem solver is a superior skill that is highly rewarded.

The Power Ups will help you excel at math and acquire math power that will serve you well for the rest of you life.

Stephen Hake
Temple City, California

Power-Up Workbook

Name ______________________________

Student Progress Chart

Fact Practice

	Power Up Facts	# Possible	Time and Score (time / # correct)									
A	40 Addition Facts	40										
B	40 Addition Facts	40										
C	40 Subtraction Facts	40										
D	40 Multiplication Facts	40										
E	40 Multiplication Facts	40										
F	40 Division Facts	40										
G	24 Fractions to Reduce	24										
H	40 Multiplication and Division Facts	40										
I	20 Improper Fractions	20										
J	20 Mixed Numbers	20										
K	Metric and Customary Conversions	30										
L	Capacity	23										
M	Percent-Decimal-Fraction Equivalents	22										
N	Measurement Facts	21										
O	Algebraic Addition of Integers	16										
P	Algebraic Addition and Subtraction of Integers	16										
Q	Algebraic Multiplication and Division of Integers	16										

Name ______________________ Time ____________

Power Up A

Use with Lesson 1

Facts

Add.

4 + 6	9 + 9	3 + 4	5 + 5	7 + 8	2 + 3	7 + 0	5 + 9	2 + 6	3 + 9
3 + 5	2 + 2	6 + 7	8 + 8	2 + 9	5 + 7	4 + 9	6 + 6	3 + 8	7 + 7
4 + 4	7 + 9	5 + 8	2 + 7	0 + 0	6 + 8	3 + 7	2 + 4	7 + 1	4 + 8
5 + 6	4 + 7	2 + 5	3 + 6	8 + 9	2 + 8	10 + 10	4 + 5	6 + 9	3 + 3

Mental Math

a.	b.	c.	d.
e.	f.	g.	h.

Problem Solving

Understand

What information am I given?

What am I asked to find or do?

Plan

How can I use the information I am given?

Which strategy should I try?

Solve

Did I follow the plan?

Did I show my work?

Did I write the answer?

Check

Did I use the correct information?

Did I do what was asked?

Is my answer reasonable?

Name ______________________ Time ____________

Power Up A

Use with Lesson 2

Facts

Add.

4 + 6	9 + 9	3 + 4	5 + 5	7 + 8	2 + 3	7 + 0	5 + 9	2 + 6	3 + 9
3 + 5	2 + 2	6 + 7	8 + 8	2 + 9	5 + 7	4 + 9	6 + 6	3 + 8	7 + 7
4 + 4	7 + 9	5 + 8	2 + 7	0 + 0	6 + 8	3 + 7	2 + 4	7 + 1	4 + 8
5 + 6	4 + 7	2 + 5	3 + 6	8 + 9	2 + 8	10 + 10	4 + 5	6 + 9	3 + 3

Mental Math

a.	b.	c.	d.
e.	f.	g.	h.

Problem Solving

Understand

What information am I given?

What am I asked to find or do?

Plan

How can I use the information I am given?

Which strategy should I try?

Solve

Did I follow the plan?

Did I show my work?

Did I write the answer?

Check

Did I use the correct information?

Did I do what was asked?

Is my answer reasonable?

Name ______________________ Time ____________

Power Up B

Use with Lesson 3

Facts

Add.

7 + 7	2 + 4	6 + 8	4 + 3	5 + 5	3 + 2	7 + 6	9 + 4	10 + 10	7 + 3
4 + 4	5 + 8	2 + 2	8 + 7	3 + 9	6 + 6	3 + 5	9 + 1	4 + 7	8 + 9
2 + 8	5 + 6	0 + 0	8 + 4	6 + 3	9 + 6	4 + 5	9 + 7	2 + 6	9 + 9
3 + 8	9 + 5	9 + 2	8 + 8	5 + 2	3 + 3	7 + 5	8 + 0	7 + 2	6 + 4

Mental Math

a.	**b.**	**c.**	**d.**
e.	**f.**	**g.**	**h.**

Problem Solving

Understand

What information am I given?

What am I asked to find or do?

Plan

How can I use the information I am given?

Which strategy should I try?

Solve

Did I follow the plan?

Did I show my work?

Did I write the answer?

Check

Did I use the correct information?

Did I do what was asked?

Is my answer reasonable?

Name ______________________ Time ____________

Power Up A

Use with Lesson 4

Facts

Add.

4 + 6	9 + 9	3 + 4	5 + 5	7 + 8	2 + 3	7 + 0	5 + 9	2 + 6	3 + 9
3 + 5	2 + 2	6 + 7	8 + 8	2 + 9	5 + 7	4 + 9	6 + 6	3 + 8	7 + 7
4 + 4	7 + 9	5 + 8	2 + 7	0 + 0	6 + 8	3 + 7	2 + 4	7 + 1	4 + 8
5 + 6	4 + 7	2 + 5	3 + 6	8 + 9	2 + 8	10 + 10	4 + 5	6 + 9	3 + 3

Mental Math

a.	b.	c.	d.
e.	f.	g.	h.

Problem Solving

Understand

What information am I given?
What am I asked to find or do?

Plan

How can I use the information I am given?
Which strategy should I try?

Solve

Did I follow the plan?
Did I show my work?
Did I write the answer?

Check

Did I use the correct information?
Did I do what was asked?
Is my answer reasonable?

Name ______________________ Time __________

Power Up B

Use with Lesson 5

Facts

Add.

7 + 7	2 + 4	6 + 8	4 + 3	5 + 5	3 + 2	7 + 6	9 + 4	10 + 10	7 + 3
4 + 4	5 + 8	2 + 2	8 + 7	3 + 9	6 + 6	3 + 5	9 + 1	4 + 7	8 + 9
2 + 8	5 + 6	0 + 0	8 + 4	6 + 3	9 + 6	4 + 5	9 + 7	2 + 6	9 + 9
3 + 8	9 + 5	9 + 2	8 + 8	5 + 2	3 + 3	7 + 5	8 + 0	7 + 2	6 + 4

Mental Math

a.	b.	c.	d.
e.	f.	g.	h.

Problem Solving

Understand
What information am I given?
What am I asked to find or do?

Plan
How can I use the information I am given?
Which strategy should I try?

Solve
Did I follow the plan?
Did I show my work?
Did I write the answer?

Check
Did I use the correct information?
Did I do what was asked?
Is my answer reasonable?

Name ______________________ Time __________

Power Up C

Use with Lesson 6

Facts

Subtract.

$8 - 5$	$10 - 4$	$12 - 6$	$6 - 3$	$8 - 4$	$14 - 7$	$20 - 10$	$11 - 5$	$7 - 4$	$13 - 6$
$7 - 2$	$15 - 8$	$9 - 7$	$17 - 9$	$10 - 5$	$8 - 1$	$16 - 7$	$6 - 0$	$12 - 3$	$9 - 5$
$13 - 5$	$11 - 7$	$14 - 8$	$10 - 7$	$5 - 3$	$15 - 6$	$6 - 4$	$10 - 8$	$18 - 9$	$15 - 7$
$12 - 4$	$11 - 2$	$16 - 8$	$9 - 9$	$13 - 4$	$11 - 8$	$9 - 6$	$14 - 9$	$8 - 6$	$12 - 5$

Mental Math

a.	b.	c.	d.
e.	f.	g.	h.

Problem Solving

Understand

What information am I given?

What am I asked to find or do?

Plan

How can I use the information I am given?

Which strategy should I try?

Solve

Did I follow the plan?

Did I show my work?

Did I write the answer?

Check

Did I use the correct information?

Did I do what was asked?

Is my answer reasonable?

Name ______________________ Time ____________

Power Up C

Use with Lesson 7

Facts

Subtract.

$8 - 5$	$10 - 4$	$12 - 6$	$6 - 3$	$8 - 4$	$14 - 7$	$20 - 10$	$11 - 5$	$7 - 4$	$13 - 6$
$7 - 2$	$15 - 8$	$9 - 7$	$17 - 9$	$10 - 5$	$8 - 1$	$16 - 7$	$6 - 0$	$12 - 3$	$9 - 5$
$13 - 5$	$11 - 7$	$14 - 8$	$10 - 7$	$5 - 3$	$15 - 6$	$6 - 4$	$10 - 8$	$18 - 9$	$15 - 7$
$12 - 4$	$11 - 2$	$16 - 8$	$9 - 9$	$13 - 4$	$11 - 8$	$9 - 6$	$14 - 9$	$8 - 6$	$12 - 5$

Mental Math

a.	b.	c.	d.
e.	f.	g.	h.

Problem Solving

Understand

What information am I given?

What am I asked to find or do?

Plan

How can I use the information I am given?

Which strategy should I try?

Solve

Did I follow the plan?

Did I show my work?

Did I write the answer?

Check

Did I use the correct information?

Did I do what was asked?

Is my answer reasonable?

Name ______________________ Time __________

Power Up A

Use with Lesson 8

Facts

Add.

4 + 6	9 + 9	3 + 4	5 + 5	7 + 8	2 + 3	7 + 0	5 + 9	2 + 6	3 + 9
3 + 5	2 + 2	6 + 7	8 + 8	2 + 9	5 + 7	4 + 9	6 + 6	3 + 8	7 + 7
4 + 4	7 + 9	5 + 8	2 + 7	0 + 0	6 + 8	3 + 7	2 + 4	7 + 1	4 + 8
5 + 6	4 + 7	2 + 5	3 + 6	8 + 9	2 + 8	10 + 10	4 + 5	6 + 9	3 + 3

Mental Math

a.	**b.**	**c.**	**d.**
e.	**f.**	**g.**	**h.**

Problem Solving

Understand

What information am I given?

What am I asked to find or do?

Plan

How can I use the information I am given?

Which strategy should I try?

Solve

Did I follow the plan?

Did I show my work?

Did I write the answer?

Check

Did I use the correct information?

Did I do what was asked?

Is my answer reasonable?

Name ________________________ Time ____________

Power Up C

Use with Lesson 9

Facts

Subtract.

8 − 5	10 − 4	12 − 6	6 − 3	8 − 4	14 − 7	20 − 10	11 − 5	7 − 4	13 − 6
7 − 2	15 − 8	9 − 7	17 − 9	10 − 5	8 − 1	16 − 7	6 − 0	12 − 3	9 − 5
13 − 5	11 − 7	14 − 8	10 − 7	5 − 3	15 − 6	6 − 4	10 − 8	18 − 9	15 − 7
12 − 4	11 − 2	16 − 8	9 − 9	13 − 4	11 − 8	9 − 6	14 − 9	8 − 6	12 − 5

Mental Math

a.	b.	c.	d.
e.	f.	g.	h.

Problem Solving

Understand
What information am I given?
What am I asked to find or do?

Plan
How can I use the information I am given?
Which strategy should I try?

Solve
Did I follow the plan?
Did I show my work?
Did I write the answer?

Check
Did I use the correct information?
Did I do what was asked?
Is my answer reasonable?

Name ______________________ Time __________

Power Up C

Use with Lesson 10

Facts

Subtract.

8 − 5	10 − 4	12 − 6	6 − 3	8 − 4	14 − 7	20 − 10	11 − 5	7 − 4	13 − 6
7 − 2	15 − 8	9 − 7	17 − 9	10 − 5	8 − 1	16 − 7	6 − 0	12 − 3	9 − 5
13 − 5	11 − 7	14 − 8	10 − 7	5 − 3	15 − 6	6 − 4	10 − 8	18 − 9	15 − 7
12 − 4	11 − 2	16 − 8	9 − 9	13 − 4	11 − 8	9 − 6	14 − 9	8 − 6	12 − 5

Mental Math

a.	b.	c.	d.
e.	f.	g.	h.

Problem Solving

Understand

What information am I given?
What am I asked to find or do?

Plan

How can I use the information I am given?
Which strategy should I try?

Solve

Did I follow the plan?
Did I show my work?
Did I write the answer?

Check

Did I use the correct information?
Did I do what was asked?
Is my answer reasonable?

Name ______________________ Time ____________

Power Up D

Use with Lesson 11

Facts Multiply.

7 × 7	4 × 6	8 × 1	2 × 2	0 × 5	6 × 3	8 × 9	5 × 8	6 × 2	10 × 10
9 × 4	2 × 5	9 × 6	7 × 3	5 × 5	7 × 2	6 × 8	3 × 5	9 × 9	5 × 4
3 × 4	6 × 5	8 × 2	4 × 4	6 × 7	8 × 8	2 × 3	7 × 4	5 × 9	3 × 8
3 × 9	7 × 8	2 × 4	5 × 7	3 × 3	9 × 7	4 × 8	0 × 0	9 × 2	6 × 6

Mental Math

a.	b.	c.	d.
e.	f.	g.	h.

Problem Solving

Understand
What information am I given?
What am I asked to find or do?

Plan
How can I use the information I am given?
Which strategy should I try?

Solve
Did I follow the plan?
Did I show my work?
Did I write the answer?

Check
Did I use the correct information?
Did I do what was asked?
Is my answer reasonable?

Name ______________________ Time ___________

Power Up D

Use with Lesson 12

Facts

Multiply.

7×7	4×6	8×1	2×2	0×5	6×3	8×9	5×8	6×2	10×10
9×4	2×5	9×6	7×3	5×5	7×2	6×8	3×5	9×9	5×4
3×4	6×5	8×2	4×4	6×7	8×8	2×3	7×4	5×9	3×8
3×9	7×8	2×4	5×7	3×3	9×7	4×8	0×0	9×2	6×6

Mental Math

a.	**b.**	**c.**	**d.**
e.	**f.**	**g.**	**h.**

Problem Solving

Understand

What information am I given?

What am I asked to find or do?

Plan

How can I use the information I am given?

Which strategy should I try?

Solve

Did I follow the plan?

Did I show my work?

Did I write the answer?

Check

Did I use the correct information?

Did I do what was asked?

Is my answer reasonable?

Name ______________________ Time ____________

Power Up A

Use with Lesson 13

Facts

Add.

4 + 6	9 + 9	3 + 4	5 + 5	7 + 8	2 + 3	7 + 0	5 + 9	2 + 6	3 + 9
3 + 5	2 + 2	6 + 7	8 + 8	2 + 9	5 + 7	4 + 9	6 + 6	3 + 8	7 + 7
4 + 4	7 + 9	5 + 8	2 + 7	0 + 0	6 + 8	3 + 7	2 + 4	7 + 1	4 + 8
5 + 6	4 + 7	2 + 5	3 + 6	8 + 9	2 + 8	10 + 10	4 + 5	6 + 9	3 + 3

Mental Math

a.	**b.**	**c.**	**d.**
e.	**f.**	**g.**	**h.**

Problem Solving

Understand

What information am I given?
What am I asked to find or do?

Plan

How can I use the information I am given?
Which strategy should I try?

Solve

Did I follow the plan?
Did I show my work?
Did I write the answer?

Check

Did I use the correct information?
Did I do what was asked?
Is my answer reasonable?

Name ____________________ Time ____________

Power Up D

Use with Lesson 14

Facts Multiply.

7 × 7	4 × 6	8 × 1	2 × 2	0 × 5	6 × 3	8 × 9	5 × 8	6 × 2	10 × 10
9 × 4	2 × 5	9 × 6	7 × 3	5 × 5	7 × 2	6 × 8	3 × 5	9 × 9	5 × 4
3 × 4	6 × 5	8 × 2	4 × 4	6 × 7	8 × 8	2 × 3	7 × 4	5 × 9	3 × 8
3 × 9	7 × 8	2 × 4	5 × 7	3 × 3	9 × 7	4 × 8	0 × 0	9 × 2	6 × 6

Mental Math

a.	b.	c.	d.
e.	f.	g.	h.

Problem Solving

Understand

What information am I given?

What am I asked to find or do?

Plan

How can I use the information I am given?

Which strategy should I try?

Solve

Did I follow the plan?

Did I show my work?

Did I write the answer?

Check

Did I use the correct information?

Did I do what was asked?

Is my answer reasonable?

Name ______________________ Time __________

Power Up C

Use with Lesson 15

Facts

Subtract.

$8 - 5$	$10 - 4$	$12 - 6$	$6 - 3$	$8 - 4$	$14 - 7$	$20 - 10$	$11 - 5$	$7 - 4$	$13 - 6$
$7 - 2$	$15 - 8$	$9 - 7$	$17 - 9$	$10 - 5$	$8 - 1$	$16 - 7$	$6 - 0$	$12 - 3$	$9 - 5$
$13 - 5$	$11 - 7$	$14 - 8$	$10 - 7$	$5 - 3$	$15 - 6$	$6 - 4$	$10 - 8$	$18 - 9$	$15 - 7$
$12 - 4$	$11 - 2$	$16 - 8$	$9 - 9$	$13 - 4$	$11 - 8$	$9 - 6$	$14 - 9$	$8 - 6$	$12 - 5$

Mental Math

a.	**b.**	**c.**	**d.**
e.	**f.**	**g.**	**h.**

Problem Solving

Understand

What information am I given?

What am I asked to find or do?

Plan

How can I use the information I am given?

Which strategy should I try?

Solve

Did I follow the plan?

Did I show my work?

Did I write the answer?

Check

Did I use the correct information?

Did I do what was asked?

Is my answer reasonable?

Name ______________ Time ______________

Power Up D

Use with Lesson 16

Facts

Multiply.

7×7	4×6	8×1	2×2	0×5	6×3	8×9	5×8	6×2	10×10
9×4	2×5	9×6	7×3	5×5	7×2	6×8	3×5	9×9	5×4
3×4	6×5	8×2	4×4	6×7	8×8	2×3	7×4	5×9	3×8
3×9	7×8	2×4	5×7	3×3	9×7	4×8	0×0	9×2	6×6

Mental Math

a.	**b.**	**c.**	**d.**
e.	**f.**	**g.**	**h.**

Problem Solving

Understand

What information am I given?

What am I asked to find or do?

Plan

How can I use the information I am given?

Which strategy should I try?

Solve

Did I follow the plan?

Did I show my work?

Did I write the answer?

Check

Did I use the correct information?

Did I do what was asked?

Is my answer reasonable?

Name ____________ Time ________

Power Up E

Use with Lesson 17

Facts Multiply.

8×8	3×9	6×7	5×2	0×0	3×8	4×6	5×8	2×9	9×9
6×1	2×6	3×3	4×5	5×5	8×6	4×2	7×7	7×4	5×3
6×9	8×4	5×9	4×3	7×8	2×2	6×5	2×7	8×9	3×6
4×4	5×7	3×2	7×9	6×6	3×7	2×8	0×7	9×4	10×10

Mental Math

a.	b.	c.	d.
e.	f.	g.	h.

Problem Solving

Understand
What information am I given?
What am I asked to find or do?

Plan
How can I use the information I am given?
Which strategy should I try?

Solve
Did I follow the plan?
Did I show my work?
Did I write the answer?

Check
Did I use the correct information?
Did I do what was asked?
Is my answer reasonable?

Name ______________________ Time __________

Power Up B

Use with Lesson 18

Facts Add.

7 + 7	2 + 4	6 + 8	4 + 3	5 + 5	3 + 2	7 + 6	9 + 4	10 + 10	7 + 3
4 + 4	5 + 8	2 + 2	8 + 7	3 + 9	6 + 6	3 + 5	9 + 1	4 + 7	8 + 9
2 + 8	5 + 6	0 + 0	8 + 4	6 + 3	9 + 6	4 + 5	9 + 7	2 + 6	9 + 9
3 + 8	9 + 5	9 + 2	8 + 8	5 + 2	3 + 3	7 + 5	8 + 0	7 + 2	6 + 4

Mental Math

a.	b.	c.	d.
e.	f.	g.	h.

Problem Solving

Understand

What information am I given?

What am I asked to find or do?

Plan

How can I use the information I am given?

Which strategy should I try?

Solve

Did I follow the plan?

Did I show my work?

Did I write the answer?

Check

Did I use the correct information?

Did I do what was asked?

Is my answer reasonable?

Name ________________________ Time ____________

Power Up D

Use with Lesson 19

Facts Multiply.

7×7	4×6	8×1	2×2	0×5	6×3	8×9	5×8	6×2	10×10
9×4	2×5	9×6	7×3	5×5	7×2	6×8	3×5	9×9	5×4
3×4	6×5	8×2	4×4	6×7	8×8	2×3	7×4	5×9	3×8
3×9	7×8	2×4	5×7	3×3	9×7	4×8	0×0	9×2	6×6

Mental Math

a.	**b.**	**c.**	**d.**
e.	**f.**	**g.**	**h.**

Problem Solving

Understand
What information am I given?
What am I asked to find or do?

Plan
How can I use the information I am given?
Which strategy should I try?

Solve
Did I follow the plan?
Did I show my work?
Did I write the answer?

Check
Did I use the correct information?
Did I do what was asked?
Is my answer reasonable?

Name ______________________ Time ____________

Power Up C

Use with Lesson 20

Facts

Subtract.

8 − 5	10 − 4	12 − 6	6 − 3	8 − 4	14 − 7	20 − 10	11 − 5	7 − 4	13 − 6
7 − 2	15 − 8	9 − 7	17 − 9	10 − 5	8 − 1	16 − 7	6 − 0	12 − 3	9 − 5
13 − 5	11 − 7	14 − 8	10 − 7	5 − 3	15 − 6	6 − 4	10 − 8	18 − 9	15 − 7
12 − 4	11 − 2	16 − 8	9 − 9	13 − 4	11 − 8	9 − 6	14 − 9	8 − 6	12 − 5

Mental Math

a.	b.	c.	d.
e.	f.	g.	h.

Problem Solving

Understand
What information am I given?
What am I asked to find or do?

Plan
How can I use the information I am given?
Which strategy should I try?

Solve
Did I follow the plan?
Did I show my work?
Did I write the answer?

Check
Did I use the correct information?
Did I do what was asked?
Is my answer reasonable?

Name ________________ Time ________

Power Up D

Use with Lesson 21

Facts

Multiply.

7×7	4×6	8×1	2×2	0×5	6×3	8×9	5×8	6×2	10×10
9×4	2×5	9×6	7×3	5×5	7×2	6×8	3×5	9×9	5×4
3×4	6×5	8×2	4×4	6×7	8×8	2×3	7×4	5×9	3×8
3×9	7×8	2×4	5×7	3×3	9×7	4×8	0×0	9×2	6×6

Mental Math

a.	b.	c.	d.
e.	f.	g.	h.

Problem Solving

Understand
What information am I given?
What am I asked to find or do?

Plan
How can I use the information I am given?
Which strategy should I try?

Solve
Did I follow the plan?
Did I show my work?
Did I write the answer?

Check
Did I use the correct information?
Did I do what was asked?
Is my answer reasonable?

Name ______________ Time ______________

Power Up C

Use with Lesson 22

Facts Subtract.

8 − 5	10 − 4	12 − 6	6 − 3	8 − 4	14 − 7	20 − 10	11 − 5	7 − 4	13 − 6
7 − 2	15 − 8	9 − 7	17 − 9	10 − 5	8 − 1	16 − 7	6 − 0	12 − 3	9 − 5
13 − 5	11 − 7	14 − 8	10 − 7	5 − 3	15 − 6	6 − 4	10 − 8	18 − 9	15 − 7
12 − 4	11 − 2	16 − 8	9 − 9	13 − 4	11 − 8	9 − 6	14 − 9	8 − 6	12 − 5

Mental Math

a.	b.	c.	d.
e.	f.	g.	h.

Problem Solving

Understand

What information am I given?

What am I asked to find or do?

Plan

How can I use the information I am given?

Which strategy should I try?

Solve

Did I follow the plan?

Did I show my work?

Did I write the answer?

Check

Did I use the correct information?

Did I do what was asked?

Is my answer reasonable?

Name ______________________ Time ____________

Facts Multiply.

7 × 7	4 × 6	8 × 1	2 × 2	0 × 5	6 × 3	8 × 9	5 × 8	6 × 2	10 × 10
9 × 4	2 × 5	9 × 6	7 × 3	5 × 5	7 × 2	6 × 8	3 × 5	9 × 9	5 × 4
3 × 4	6 × 5	8 × 2	4 × 4	6 × 7	8 × 8	2 × 3	7 × 4	5 × 9	3 × 8
3 × 9	7 × 8	2 × 4	5 × 7	3 × 3	9 × 7	4 × 8	0 × 0	9 × 2	6 × 6

Mental Math

a.	b.	c.	d.
e.	f.	g.	h.

Problem Solving

Understand
What information am I given?
What am I asked to find or do?

Plan
How can I use the information I am given?
Which strategy should I try?

Solve
Did I follow the plan?
Did I show my work?
Did I write the answer?

Check
Did I use the correct information?
Did I do what was asked?
Is my answer reasonable?

Name ______________________ Time __________

Power Up C

Use with Lesson 24

Facts

Subtract.

8 − 5	10 − 4	12 − 6	6 − 3	8 − 4	14 − 7	20 − 10	11 − 5	7 − 4	13 − 6
7 − 2	15 − 8	9 − 7	17 − 9	10 − 5	8 − 1	16 − 7	6 − 0	12 − 3	9 − 5
13 − 5	11 − 7	14 − 8	10 − 7	5 − 3	15 − 6	6 − 4	10 − 8	18 − 9	15 − 7
12 − 4	11 − 2	16 − 8	9 − 9	13 − 4	11 − 8	9 − 6	14 − 9	8 − 6	12 − 5

Mental Math

a.	b.	c.	d.
e.	f.	g.	h.

Problem Solving

Understand
What information am I given?
What am I asked to find or do?

Plan
How can I use the information I am given?
Which strategy should I try?

Solve
Did I follow the plan?
Did I show my work?
Did I write the answer?

Check
Did I use the correct information?
Did I do what was asked?
Is my answer reasonable?

Name ____________________ Time __________

Power Up F

Use with Lesson 25

Facts

Divide.

$7\overline{)49}$	$9\overline{)27}$	$5\overline{)25}$	$4\overline{)12}$	$6\overline{)36}$	$7\overline{)21}$	$10\overline{)100}$	$5\overline{)10}$	$4\overline{)0}$	$4\overline{)16}$
$8\overline{)72}$	$4\overline{)28}$	$2\overline{)14}$	$7\overline{)35}$	$5\overline{)40}$	$2\overline{)8}$	$8\overline{)8}$	$3\overline{)9}$	$8\overline{)24}$	$4\overline{)24}$
$6\overline{)54}$	$3\overline{)18}$	$8\overline{)56}$	$3\overline{)6}$	$8\overline{)48}$	$5\overline{)20}$	$2\overline{)16}$	$7\overline{)63}$	$6\overline{)12}$	$1\overline{)6}$
$4\overline{)32}$	$9\overline{)45}$	$2\overline{)18}$	$8\overline{)64}$	$6\overline{)30}$	$5\overline{)15}$	$6\overline{)42}$	$3\overline{)24}$	$9\overline{)81}$	$4\overline{)36}$

Mental Math

a.	**b.**	**c.**	**d.**
e.	**f.**	**g.**	**h.**

Problem Solving

Understand

What information am I given?

What am I asked to find or do?

Plan

How can I use the information I am given?

Which strategy should I try?

Solve

Did I follow the plan?

Did I show my work?

Did I write the answer?

Check

Did I use the correct information?

Did I do what was asked?

Is my answer reasonable?

Name ____________________ Time __________

Power Up C

Use with Lesson 26

Facts Subtract.

8 − 5	10 − 4	12 − 6	6 − 3	8 − 4	14 − 7	20 − 10	11 − 5	7 − 4	13 − 6
7 − 2	15 − 8	9 − 7	17 − 9	10 − 5	8 − 1	16 − 7	6 − 0	12 − 3	9 − 5
13 − 5	11 − 7	14 − 8	10 − 7	5 − 3	15 − 6	6 − 4	10 − 8	18 − 9	15 − 7
12 − 4	11 − 2	16 − 8	9 − 9	13 − 4	11 − 8	9 − 6	14 − 9	8 − 6	12 − 5

Mental Math

a.	b.	c.	d.
e.	f.	g.	h.

Problem Solving

Understand

What information am I given?

What am I asked to find or do?

Plan

How can I use the information I am given?

Which strategy should I try?

Solve

Did I follow the plan?

Did I show my work?

Did I write the answer?

Check

Did I use the correct information?

Did I do what was asked?

Is my answer reasonable?

Name ________________ Time ________

Power Up E

Use with Lesson 27

Facts

Multiply.

8×8	3×9	6×7	5×2	0×0	3×8	4×6	5×8	2×9	9×9
6×1	2×6	3×3	4×5	5×5	8×6	4×2	7×7	7×4	5×3
6×9	8×4	5×9	4×3	7×8	2×2	6×5	2×7	8×9	3×6
4×4	5×7	3×2	7×9	6×6	3×7	2×8	0×7	9×4	10×10

Mental Math

a.	**b.**	**c.**	**d.**
e.	**f.**	**g.**	**h.**

Problem Solving

Understand

What information am I given?

What am I asked to find or do?

Plan

How can I use the information I am given?

Which strategy should I try?

Solve

Did I follow the plan?

Did I show my work?

Did I write the answer?

Check

Did I use the correct information?

Did I do what was asked?

Is my answer reasonable?

Name ________________ Time ________

Power Up F

Use with Lesson 28

Facts

Divide.

$7\overline{)49}$	$9\overline{)27}$	$5\overline{)25}$	$4\overline{)12}$	$6\overline{)36}$	$7\overline{)21}$	$10\overline{)100}$	$5\overline{)10}$	$4\overline{)0}$	$4\overline{)16}$
$8\overline{)72}$	$4\overline{)28}$	$2\overline{)14}$	$7\overline{)35}$	$5\overline{)40}$	$2\overline{)8}$	$8\overline{)8}$	$3\overline{)9}$	$8\overline{)24}$	$4\overline{)24}$
$6\overline{)54}$	$3\overline{)18}$	$8\overline{)56}$	$3\overline{)6}$	$8\overline{)48}$	$5\overline{)20}$	$2\overline{)16}$	$7\overline{)63}$	$6\overline{)12}$	$1\overline{)6}$
$4\overline{)32}$	$9\overline{)45}$	$2\overline{)18}$	$8\overline{)64}$	$6\overline{)30}$	$5\overline{)15}$	$6\overline{)42}$	$3\overline{)24}$	$9\overline{)81}$	$4\overline{)36}$

Mental Math

a.	**b.**	**c.**	**d.**
e.	**f.**	**g.**	**h.**

Problem Solving

Understand

What information am I given?

What am I asked to find or do?

Plan

How can I use the information I am given?

Which strategy should I try?

Solve

Did I follow the plan?

Did I show my work?

Did I write the answer?

Check

Did I use the correct information?

Did I do what was asked?

Is my answer reasonable?

Name ______________________ Time ____________

Power Up B

Use with Lesson 29

Facts

Add.

7 + 7	2 + 4	6 + 8	4 + 3	5 + 5	3 + 2	7 + 6	9 + 4	10 + 10	7 + 3
4 + 4	5 + 8	2 + 2	8 + 7	3 + 9	6 + 6	3 + 5	9 + 1	4 + 7	8 + 9
2 + 8	5 + 6	0 + 0	8 + 4	6 + 3	0 + 6	4 + 5	9 + 7	2 + 6	9 + 9
3 + 8	9 + 5	9 + 2	8 + 8	5 + 2	3 + 3	7 + 5	8 + 0	7 + 2	6 + 4

Mental Math

a.	**b.**	**c.**	**d.**
e.	**f.**	**g.**	**h.**

Problem Solving

Understand
What information am I given?
What am I asked to find or do?

Plan
How can I use the information I am given?
Which strategy should I try?

Solve
Did I follow the plan?
Did I show my work?
Did I write the answer?

Check
Did I use the correct information?
Did I do what was asked?
Is my answer reasonable?

Name ____________________ Time ______________

Power Up E

Use with Lesson 30

Facts

Multiply.

8 × 8	3 × 9	6 × 7	5 × 2	0 × 0	3 × 8	4 × 6	5 × 8	2 × 9	9 × 9
6 × 1	2 × 6	3 × 3	4 × 5	5 × 5	8 × 6	4 × 2	7 × 7	7 × 4	5 × 3
6 × 9	8 × 4	5 × 9	4 × 3	7 × 8	2 × 2	6 × 5	2 × 7	8 × 9	3 × 6
4 × 4	5 × 7	3 × 2	7 × 9	6 × 6	3 × 7	2 × 8	0 × 7	9 × 4	10 × 10

Mental Math

a.	b.	c.	d.
e.	f.	g.	h.

Problem Solving

Understand
What information am I given?
What am I asked to find or do?

Plan
How can I use the information I am given?
Which strategy should I try?

Solve
Did I follow the plan?
Did I show my work?
Did I write the answer?

Check
Did I use the correct information?
Did I do what was asked?
Is my answer reasonable?

Name ____________________ Time __________

Power Up F

Use with Lesson 31

Facts Divide.

$7\overline{)49}$	$9\overline{)27}$	$5\overline{)25}$	$4\overline{)12}$	$6\overline{)36}$	$7\overline{)21}$	$10\overline{)100}$	$5\overline{)10}$	$4\overline{)0}$	$4\overline{)16}$
$8\overline{)72}$	$4\overline{)28}$	$2\overline{)14}$	$7\overline{)35}$	$5\overline{)40}$	$2\overline{)8}$	$8\overline{)8}$	$3\overline{)9}$	$8\overline{)24}$	$4\overline{)24}$
$6\overline{)54}$	$3\overline{)18}$	$8\overline{)56}$	$3\overline{)6}$	$8\overline{)48}$	$5\overline{)20}$	$2\overline{)16}$	$7\overline{)63}$	$6\overline{)12}$	$1\overline{)6}$
$4\overline{)32}$	$9\overline{)45}$	$2\overline{)18}$	$8\overline{)64}$	$6\overline{)30}$	$5\overline{)15}$	$6\overline{)42}$	$3\overline{)24}$	$9\overline{)81}$	$4\overline{)36}$

Mental Math

a.	**b.**	**c.**	**d.**
e.	**f.**	**g.**	**h.**

Problem Solving

Understand

What information am I given?

What am I asked to find or do?

Plan

How can I use the information I am given?

Which strategy should I try?

Solve

Did I follow the plan?

Did I show my work?

Did I write the answer?

Check

Did I use the correct information?

Did I do what was asked?

Is my answer reasonable?

Name ______________________ Time ____________

Power Up G

Use with Lesson 32

Facts Reduce each fraction to lowest terms.

$\frac{2}{8} =$	$\frac{4}{6} =$	$\frac{6}{10} =$	$\frac{2}{4} =$	$\frac{5}{100} =$	$\frac{9}{12} =$
$\frac{4}{10} =$	$\frac{4}{12} =$	$\frac{2}{10} =$	$\frac{3}{6} =$	$\frac{25}{100} =$	$\frac{3}{12} =$
$\frac{4}{16} =$	$\frac{3}{9} =$	$\frac{6}{9} =$	$\frac{4}{8} =$	$\frac{2}{12} =$	$\frac{6}{12} =$
$\frac{8}{16} =$	$\frac{2}{6} =$	$\frac{8}{12} =$	$\frac{6}{8} =$	$\frac{5}{10} =$	$\frac{75}{100} =$

Mental Math

a.	**b.**	**c.**	**d.**
e.	**f.**	**g.**	**h.**

Problem Solving

Understand

What information am I given?

What am I asked to find or do?

Plan

How can I use the information I am given?

Which strategy should I try?

Solve

Did I follow the plan?

Did I show my work?

Did I write the answer?

Check

Did I use the correct information?

Did I do what was asked?

Is my answer reasonable?

Name ______________________ Time ____________

Power Up G

Use with Lesson 33

Facts Reduce each fraction to lowest terms.

$\frac{2}{8} =$	$\frac{4}{6} =$	$\frac{6}{10} =$	$\frac{2}{4} =$	$\frac{5}{100} =$	$\frac{9}{12} =$
$\frac{4}{10} =$	$\frac{4}{12} =$	$\frac{2}{10} =$	$\frac{3}{6} =$	$\frac{25}{100} =$	$\frac{3}{12} =$
$\frac{4}{16} =$	$\frac{3}{9} =$	$\frac{6}{9} =$	$\frac{4}{8} =$	$\frac{2}{12} =$	$\frac{6}{12} =$
$\frac{8}{16} =$	$\frac{2}{6} =$	$\frac{8}{12} =$	$\frac{6}{8} =$	$\frac{5}{10} =$	$\frac{75}{100} =$

Mental Math

a.	b.	c.	d.
e.	f.	g.	h.

Problem Solving

Understand

What information am I given?

What am I asked to find or do?

Plan

How can I use the information I am given?

Which strategy should I try?

Solve

Did I follow the plan?

Did I show my work?

Did I write the answer?

Check

Did I use the correct information?

Did I do what was asked?

Is my answer reasonable?

Name ____________________ Time __________

Power Up D

Use with Lesson 34

Facts Multiply.

7 × 7	4 × 6	8 × 1	2 × 2	0 × 5	6 × 3	8 × 9	5 × 8	6 × 2	10 × 10
9 × 4	2 × 5	9 × 6	7 × 3	5 × 5	7 × 2	6 × 8	3 × 5	9 × 9	5 × 4
3 × 4	6 × 5	8 × 2	4 × 4	6 × 7	8 × 8	2 × 3	7 × 4	5 × 9	3 × 8
3 × 9	7 × 8	2 × 4	5 × 7	3 × 3	9 × 7	4 × 8	0 × 0	9 × 2	6 × 6

Mental Math

a.	b.	c.	d.
e.	f.	g.	h.

Problem Solving

Understand
What information am I given?
What am I asked to find or do?

Plan
How can I use the information I am given?
Which strategy should I try?

Solve
Did I follow the plan?
Did I show my work?
Did I write the answer?

Check
Did I use the correct information?
Did I do what was asked?
Is my answer reasonable?

Name ______________________ Time ____________

Power Up G

Use with Lesson 35

Facts Reduce each fraction to lowest terms.

$\frac{2}{8} =$	$\frac{4}{6} =$	$\frac{6}{10} =$	$\frac{2}{4} =$	$\frac{5}{100} =$	$\frac{9}{12} =$
$\frac{4}{10} =$	$\frac{4}{12} =$	$\frac{2}{10} =$	$\frac{3}{6} =$	$\frac{25}{100} =$	$\frac{3}{12} =$
$\frac{4}{16} =$	$\frac{3}{9} =$	$\frac{6}{9} =$	$\frac{4}{8} =$	$\frac{2}{12} =$	$\frac{6}{12} =$
$\frac{8}{16} =$	$\frac{2}{6} =$	$\frac{8}{12} =$	$\frac{6}{8} =$	$\frac{5}{10} =$	$\frac{75}{100} =$

Mental Math

a.	**b.**	**c.**	**d.**
e.	**f.**	**g.**	**h.**

Problem Solving

Understand

What information am I given?

What am I asked to find or do?

Plan

How can I use the information I am given?

Which strategy should I try?

Solve

Did I follow the plan?

Did I show my work?

Did I write the answer?

Check

Did I use the correct information?

Did I do what was asked?

Is my answer reasonable?

Name ______________________ Time __________

Power Up F

Use with Lesson 36

Facts Divide.

$7\overline{)49}$	$9\overline{)27}$	$5\overline{)25}$	$4\overline{)12}$	$6\overline{)36}$	$7\overline{)21}$	$10\overline{)100}$	$5\overline{)10}$	$4\overline{)0}$	$4\overline{)16}$
$8\overline{)72}$	$4\overline{)28}$	$2\overline{)14}$	$7\overline{)35}$	$5\overline{)40}$	$2\overline{)8}$	$8\overline{)8}$	$3\overline{)9}$	$8\overline{)24}$	$4\overline{)24}$
$6\overline{)54}$	$3\overline{)18}$	$8\overline{)56}$	$3\overline{)6}$	$8\overline{)48}$	$5\overline{)20}$	$2\overline{)16}$	$7\overline{)63}$	$6\overline{)12}$	$1\overline{)6}$
$4\overline{)32}$	$9\overline{)45}$	$2\overline{)18}$	$8\overline{)64}$	$6\overline{)30}$	$5\overline{)15}$	$6\overline{)42}$	$3\overline{)24}$	$9\overline{)81}$	$4\overline{)36}$

Mental Math

a.	**b.**	**c.**	**d.**
e.	**f.**	**g.**	**h.**

Problem Solving

Understand
What information am I given?
What am I asked to find or do?

Plan
How can I use the information I am given?
Which strategy should I try?

Solve
Did I follow the plan?
Did I show my work?
Did I write the answer?

Check
Did I use the correct information?
Did I do what was asked?
Is my answer reasonable?

Name ______________________ Time __________

Power Up G

Use with Lesson 37

Facts Reduce each fraction to lowest terms.

$\frac{2}{8} =$	$\frac{4}{6} =$	$\frac{6}{10} =$	$\frac{2}{4} =$	$\frac{5}{100} =$	$\frac{9}{12} =$
$\frac{4}{10} =$	$\frac{4}{12} =$	$\frac{2}{10} =$	$\frac{3}{6} =$	$\frac{25}{100} =$	$\frac{3}{12} =$
$\frac{4}{16} =$	$\frac{3}{9} =$	$\frac{6}{9} =$	$\frac{4}{8} =$	$\frac{2}{12} =$	$\frac{6}{12} =$
$\frac{8}{16} =$	$\frac{2}{6} =$	$\frac{8}{12} =$	$\frac{6}{8} =$	$\frac{5}{10} =$	$\frac{75}{100} =$

Mental Math

a.	**b.**	**c.**	**d.**
e.	**f.**	**g.**	**h.**

Problem Solving

Understand

What information am I given?

What am I asked to find or do?

Plan

How can I use the information I am given?

Which strategy should I try?

Solve

Did I follow the plan?

Did I show my work?

Did I write the answer?

Check

Did I use the correct information?

Did I do what was asked?

Is my answer reasonable?

Name ______________________ Time __________

Power Up A

Use with Lesson 38

Facts

Add.

$4 + 6$	$9 + 9$	$3 + 4$	$5 + 5$	$7 + 8$	$2 + 3$	$7 + 0$	$5 + 9$	$2 + 6$	$3 + 9$
$3 + 5$	$2 + 2$	$6 + 7$	$8 + 8$	$2 + 9$	$5 + 7$	$4 + 9$	$6 + 6$	$3 + 8$	$7 + 7$
$4 + 4$	$7 + 9$	$5 + 8$	$2 + 7$	$0 + 0$	$6 + 8$	$3 + 7$	$2 + 4$	$7 + 1$	$4 + 8$
$5 + 6$	$4 + 7$	$2 + 5$	$3 + 6$	$8 + 9$	$2 + 8$	$10 + 10$	$4 + 5$	$6 + 9$	$3 + 3$

Mental Math

a.	**b.**	**c.**	**d.**
e.	**f.**	**g.**	**h.**

Problem Solving

Understand

What information am I given?

What am I asked to find or do?

Plan

How can I use the information I am given?

Which strategy should I try?

Solve

Did I follow the plan?

Did I show my work?

Did I write the answer?

Check

Did I use the correct information?

Did I do what was asked?

Is my answer reasonable?

Name ________________________ Time ____________

Power Up G

Use with Lesson 39

Facts Reduce each fraction to lowest terms.

$\frac{2}{8} =$	$\frac{4}{6} =$	$\frac{6}{10} =$	$\frac{2}{4} =$	$\frac{5}{100} =$	$\frac{9}{12} =$
$\frac{4}{10} =$	$\frac{4}{12} =$	$\frac{2}{10} =$	$\frac{3}{6} =$	$\frac{25}{100} =$	$\frac{3}{12} =$
$\frac{4}{16} =$	$\frac{3}{9} =$	$\frac{6}{9} =$	$\frac{4}{8} =$	$\frac{2}{12} =$	$\frac{6}{12} =$
$\frac{8}{16} =$	$\frac{2}{6} =$	$\frac{8}{12} =$	$\frac{6}{8} =$	$\frac{5}{10} =$	$\frac{75}{100} =$

Mental Math

a.	**b.**	**c.**	**d.**
e.	**f.**	**g.**	**h.**

Problem Solving

Understand
What information am I given?
What am I asked to find or do?

Plan
How can I use the information I am given?
Which strategy should I try?

Solve
Did I follow the plan?
Did I show my work?
Did I write the answer?

Check
Did I use the correct information?
Did I do what was asked?
Is my answer reasonable?

Name ______________ Time ________

Power Up D

Use with Lesson 40

Facts

Multiply.

7×7	4×6	8×1	2×2	0×5	6×3	8×9	5×8	6×2	10×10
9×4	2×5	9×6	7×3	5×5	7×2	6×8	3×5	9×9	5×4
3×4	6×5	8×2	4×4	6×7	8×8	2×3	7×4	5×9	3×8
3×9	7×8	2×4	5×7	3×3	9×7	4×8	0×0	9×2	6×6

Mental Math

a.	**b.**	**c.**	**d.**
e.	**f.**	**g.**	**h.**

Problem Solving

Understand
What information am I given?
What am I asked to find or do?

Plan
How can I use the information I am given?
Which strategy should I try?

Solve
Did I follow the plan?
Did I show my work?
Did I write the answer?

Check
Did I use the correct information?
Did I do what was asked?
Is my answer reasonable?

Name ________________ Time ________

Power Up G

Use with Lesson 41

Facts Reduce each fraction to lowest terms.

$\frac{2}{8}$ =	$\frac{4}{6}$ =	$\frac{6}{10}$ =	$\frac{2}{4}$ =	$\frac{5}{100}$ =	$\frac{9}{12}$ =
$\frac{4}{10}$ =	$\frac{4}{12}$ =	$\frac{2}{10}$ =	$\frac{3}{6}$ =	$\frac{25}{100}$ =	$\frac{3}{12}$ =
$\frac{4}{16}$ =	$\frac{3}{9}$ =	$\frac{6}{9}$ =	$\frac{4}{8}$ =	$\frac{2}{12}$ =	$\frac{6}{12}$ =
$\frac{8}{16}$ =	$\frac{2}{6}$ =	$\frac{8}{12}$ =	$\frac{6}{8}$ =	$\frac{5}{10}$ =	$\frac{75}{100}$ =

Mental Math

a.	**b.**	**c.**	**d.**
e.	**f.**	**g.**	**h.**

Problem Solving

Understand
What information am I given?
What am I asked to find or do?

Plan
How can I use the information I am given?
Which strategy should I try?

Solve
Did I follow the plan?
Did I show my work?
Did I write the answer?

Check
Did I use the correct information?
Did I do what was asked?
Is my answer reasonable?

Name ______________________ Time ________

Power Up G

Use with Lesson 42

Facts Reduce each fraction to lowest terms.

$\frac{2}{8}$ =	$\frac{4}{6}$ =	$\frac{6}{10}$ =	$\frac{2}{4}$ =	$\frac{5}{100}$ =	$\frac{9}{12}$ =
$\frac{4}{10}$ =	$\frac{4}{12}$ =	$\frac{2}{10}$ =	$\frac{3}{6}$ =	$\frac{25}{100}$ =	$\frac{3}{12}$ =
$\frac{4}{16}$ =	$\frac{3}{9}$ =	$\frac{6}{9}$ =	$\frac{4}{8}$ =	$\frac{2}{12}$ =	$\frac{6}{12}$ =
$\frac{8}{16}$ =	$\frac{2}{6}$ =	$\frac{8}{12}$ =	$\frac{6}{8}$ =	$\frac{5}{10}$ =	$\frac{75}{100}$ =

Mental Math

a.	**b.**	**c.**	**d.**
e.	**f.**	**g.**	**h.**

Problem Solving

Understand

What information am I given?

What am I asked to find or do?

Plan

How can I use the information I am given?

Which strategy should I try?

Solve

Did I follow the plan?

Did I show my work?

Did I write the answer?

Check

Did I use the correct information?

Did I do what was asked?

Is my answer reasonable?

Name ____________________ Time __________

Power Up C

Use with Lesson 43

Facts Subtract.

$8 - 5$	$10 - 4$	$12 - 6$	$6 - 3$	$8 - 4$	$14 - 7$	$20 - 10$	$11 - 5$	$7 - 4$	$13 - 6$
$7 - 2$	$15 - 8$	$9 - 7$	$17 - 9$	$10 - 5$	$8 - 1$	$16 - 7$	$6 - 0$	$12 - 3$	$9 - 5$
$13 - 5$	$11 - 7$	$14 - 8$	$10 - 7$	$5 - 3$	$15 - 6$	$6 - 4$	$10 - 8$	$18 - 9$	$15 - 7$
$12 - 4$	$11 - 2$	$16 - 8$	$9 - 9$	$13 - 4$	$11 - 8$	$9 - 6$	$14 - 9$	$8 - 6$	$12 - 5$

Mental Math

a.	**b.**	**c.**	**d.**
e.	**f.**	**g.**	**h.**

Problem Solving

Understand
What information am I given?
What am I asked to find or do?

Plan
How can I use the information I am given?
Which strategy should I try?

Solve
Did I follow the plan?
Did I show my work?
Did I write the answer?

Check
Did I use the correct information?
Did I do what was asked?
Is my answer reasonable?

Name ______________________ Time __________

Power Up G

Use with Lesson 44

Facts Reduce each fraction to lowest terms.

$\frac{2}{8} =$	$\frac{4}{6} =$	$\frac{6}{10} =$	$\frac{2}{4} =$	$\frac{5}{100} =$	$\frac{9}{12} =$
$\frac{4}{10} =$	$\frac{4}{12} =$	$\frac{2}{10} =$	$\frac{3}{6} =$	$\frac{25}{100} =$	$\frac{3}{12} =$
$\frac{4}{16} =$	$\frac{3}{9} =$	$\frac{6}{9} =$	$\frac{4}{8} =$	$\frac{2}{12} =$	$\frac{6}{12} =$
$\frac{8}{16} =$	$\frac{2}{6} =$	$\frac{8}{12} =$	$\frac{6}{8} =$	$\frac{5}{10} =$	$\frac{75}{100} =$

Mental Math

a.	**b.**	**c.**	**d.**
e.	**f.**	**g.**	**h.**

Problem Solving

Understand
What information am I given?
What am I asked to find or do?

Plan
How can I use the information I am given?
Which strategy should I try?

Solve
Did I follow the plan?
Did I show my work?
Did I write the answer?

Check
Did I use the correct information?
Did I do what was asked?
Is my answer reasonable?

Name ____________________ Time ____________

Power Up F

Use with Lesson 45

Facts Divide.

7)49	9)27	5)25	4)12	6)36	7)21	10)100	5)10	4)0	4)16
8)72	4)28	2)14	7)35	5)40	2)8	8)8	3)9	8)24	4)24
6)54	3)18	8)56	3)6	8)48	5)20	2)16	7)63	6)12	1)6
4)32	9)45	2)18	8)64	6)30	5)15	6)42	3)24	9)81	4)36

Mental Math

a.	**b.**	**c.**	**d.**
e.	**f.**	**g.**	**h.**

Problem Solving

Understand

What information am I given?

What am I asked to find or do?

Plan

How can I use the information I am given?

Which strategy should I try?

Solve

Did I follow the plan?

Did I show my work?

Did I write the answer?

Check

Did I use the correct information?

Did I do what was asked?

Is my answer reasonable?

Name ______________________ Time ____________

Power Up G

Use with Lesson 46

Facts Reduce each fraction to lowest terms.

$\frac{2}{8} =$	$\frac{4}{6} =$	$\frac{6}{10} =$	$\frac{2}{4} =$	$\frac{5}{100} =$	$\frac{9}{12} =$
$\frac{4}{10} =$	$\frac{4}{12} =$	$\frac{2}{10} =$	$\frac{3}{6} =$	$\frac{25}{100} =$	$\frac{3}{12} =$
$\frac{4}{16} =$	$\frac{3}{9} =$	$\frac{6}{9} =$	$\frac{4}{8} =$	$\frac{2}{12} =$	$\frac{6}{12} =$
$\frac{8}{16} =$	$\frac{2}{6} =$	$\frac{8}{12} =$	$\frac{6}{8} =$	$\frac{5}{10} =$	$\frac{75}{100} =$

Mental Math

a.	**b.**	**c.**	**d.**
e.	**f.**	**g.**	**h.**

Problem Solving

Understand
What information am I given?
What am I asked to find or do?

Plan
How can I use the information I am given?
Which strategy should I try?

Solve
Did I follow the plan?
Did I show my work?
Did I write the answer?

Check
Did I use the correct information?
Did I do what was asked?
Is my answer reasonable?

Name ______________________ Time ____________

Power Up E

Use with Lesson 47

Facts Multiply.

8×8	3×9	6×7	5×2	0×0	3×8	4×6	5×8	2×9	9×9
6×1	2×6	3×3	4×5	5×5	8×6	4×2	7×7	7×4	5×3
6×9	8×4	5×9	4×3	7×8	2×2	6×5	2×7	8×9	3×6
4×4	5×7	3×2	7×9	6×6	3×7	2×8	0×7	9×4	10×10

Mental Math

a.	**b.**	**c.**	**d.**
e.	**f.**	**g.**	**h.**

Problem Solving

Understand
What information am I given?
What am I asked to find or do?

Plan
How can I use the information I am given?
Which strategy should I try?

Solve
Did I follow the plan?
Did I show my work?
Did I write the answer?

Check
Did I use the correct information?
Did I do what was asked?
Is my answer reasonable?

Name ____________________ Time ____________

Power Up G

Use with Lesson 48

Facts Reduce each fraction to lowest terms.

$\frac{2}{8} =$	$\frac{4}{6} =$	$\frac{6}{10} =$	$\frac{2}{4} =$	$\frac{5}{100} =$	$\frac{9}{12} =$
$\frac{4}{10} =$	$\frac{4}{12} =$	$\frac{2}{10} =$	$\frac{3}{6} =$	$\frac{25}{100} =$	$\frac{3}{12} =$
$\frac{4}{16} =$	$\frac{3}{9} =$	$\frac{6}{9} =$	$\frac{4}{8} =$	$\frac{2}{12} =$	$\frac{6}{12} =$
$\frac{8}{16} =$	$\frac{2}{6} =$	$\frac{8}{12} =$	$\frac{6}{8} =$	$\frac{5}{10} =$	$\frac{75}{100} =$

Mental Math

a.	**b.**	**c.**	**d.**
e.	**f.**	**g.**	**h.**

Problem Solving

Understand
What information am I given?
What am I asked to find or do?

Plan
How can I use the information I am given?
Which strategy should I try?

Solve
Did I follow the plan?
Did I show my work?
Did I write the answer?

Check
Did I use the correct information?
Did I do what was asked?
Is my answer reasonable?

Name ____________________ Time __________

Power Up H

Use with Lesson 49

Facts

Multiply or divide as indicated.

4×9	$4\overline{)16}$	6×8	$3\overline{)12}$	5×7	$4\overline{)32}$	3×9	$9\overline{)81}$	6×2	$8\overline{)64}$
9×7	$8\overline{)40}$	2×4	$6\overline{)42}$	5×5	$7\overline{)14}$	7×7	$8\overline{)8}$	3×3	$6\overline{)0}$
7×3	$2\overline{)10}$	10×10	$3\overline{)24}$	4×5	$9\overline{)54}$	9×1	$3\overline{)6}$	7×4	$7\overline{)56}$
6×6	$2\overline{)18}$	3×5	$5\overline{)30}$	2×2	$6\overline{)18}$	9×5	$6\overline{)24}$	2×8	$9\overline{)72}$

Mental Math

a.	**b.**	**c.**	**d.**
e.	**f.**	**g.**	**h.**

Problem Solving

Understand
What information am I given?
What am I asked to find or do?

Plan
How can I use the information I am given?
Which strategy should I try?

Solve
Did I follow the plan?
Did I show my work?
Did I write the answer?

Check
Did I use the correct information?
Did I do what was asked?
Is my answer reasonable?

Name ______________________ Time ____________

Power Up G

Use with Lesson 50

Facts Reduce each fraction to lowest terms.

$\frac{2}{8}$ =	$\frac{4}{6}$ =	$\frac{6}{10}$ =	$\frac{2}{4}$ =	$\frac{5}{100}$ =	$\frac{9}{12}$ =
$\frac{4}{10}$ =	$\frac{4}{12}$ =	$\frac{2}{10}$ =	$\frac{3}{6}$ =	$\frac{25}{100}$ =	$\frac{3}{12}$ =
$\frac{4}{16}$ =	$\frac{3}{9}$ =	$\frac{6}{9}$ =	$\frac{4}{8}$ =	$\frac{2}{12}$ =	$\frac{6}{12}$ =
$\frac{8}{16}$ =	$\frac{2}{6}$ =	$\frac{8}{12}$ =	$\frac{6}{8}$ =	$\frac{5}{10}$ =	$\frac{75}{100}$ =

Mental Math

a.	**b.**	**c.**	**d.**
e.	**f.**	**g.**	**h.**

Problem Solving

Understand

What information am I given?

What am I asked to find or do?

Plan

How can I use the information I am given?

Which strategy should I try?

Solve

Did I follow the plan?

Did I show my work?

Did I write the answer?

Check

Did I use the correct information?

Did I do what was asked?

Is my answer reasonable?

Name ______________________ Time ____________

Power Up H

Use with Lesson 51

Facts Multiply or divide as indicated.

4×9	$16 \div 4$	6×8	$12 \div 3$	5×7	$32 \div 4$	3×9	$81 \div 9$	6×2	$64 \div 8$
9×7	$40 \div 8$	2×4	$42 \div 6$	5×5	$14 \div 7$	7×7	$8 \div 8$	3×3	$0 \div 6$
7×3	$10 \div 2$	10×10	$24 \div 3$	4×5	$54 \div 9$	9×1	$6 \div 3$	7×4	$56 \div 7$
6×6	$18 \div 2$	3×5	$30 \div 5$	2×2	$18 \div 6$	9×5	$24 \div 6$	2×8	$72 \div 9$

Mental Math

a.	b.	c.	d.
e.	f.	g.	h.

Problem Solving

Understand
What information am I given?
What am I asked to find or do?

Plan
How can I use the information I am given?
Which strategy should I try?

Solve
Did I follow the plan?
Did I show my work?
Did I write the answer?

Check
Did I use the correct information?
Did I do what was asked?
Is my answer reasonable?

Name ____________________ Time __________

Power Up G

Use with Lesson 52

Facts Reduce each fraction to lowest terms.

$\frac{2}{8} =$	$\frac{4}{6} =$	$\frac{6}{10} =$	$\frac{2}{4} =$	$\frac{5}{100} =$	$\frac{9}{12} =$
$\frac{4}{10} =$	$\frac{4}{12} =$	$\frac{2}{10} =$	$\frac{3}{6} =$	$\frac{25}{100} =$	$\frac{3}{12} =$
$\frac{4}{16} =$	$\frac{3}{9} =$	$\frac{6}{9} =$	$\frac{4}{8} =$	$\frac{2}{12} =$	$\frac{6}{12} =$
$\frac{8}{16} =$	$\frac{2}{6} =$	$\frac{8}{12} =$	$\frac{6}{8} =$	$\frac{5}{10} =$	$\frac{75}{100} =$

Mental Math

a.	**b.**	**c.**	**d.**
e.	**f.**	**g.**	**h.**

Problem Solving

Understand

What information am I given?

What am I asked to find or do?

Plan

How can I use the information I am given?

Which strategy should I try?

Solve

Did I follow the plan?

Did I show my work?

Did I write the answer?

Check

Did I use the correct information?

Did I do what was asked?

Is my answer reasonable?

Name ______________________ Time ____________

Power Up

Use with Lesson 53

Facts Multiply.

7×7	4×6	8×1	2×2	0×5	6×3	8×9	5×8	6×2	10×10
9×4	2×5	9×6	7×3	5×5	7×2	6×8	3×5	9×9	5×4
3×4	6×5	8×2	4×4	6×7	8×8	2×3	7×4	5×9	3×8
3×9	7×8	2×4	5×7	3×3	9×7	4×8	0×0	9×2	6×6

Mental Math

a.	**b.**	**c.**	**d.**
e.	**f.**	**g.**	**h.**

Problem Solving

Understand
What information am I given?
What am I asked to find or do?

Plan
How can I use the information I am given?
Which strategy should I try?

Solve
Did I follow the plan?
Did I show my work?
Did I write the answer?

Check
Did I use the correct information?
Did I do what was asked?
Is my answer reasonable?

Name ______________________ Time __________

Power Up G

Use with Lesson 54

Facts

Reduce each fraction to lowest terms.

$\frac{2}{8}$ =	$\frac{4}{6}$ =	$\frac{6}{10}$ =	$\frac{2}{4}$ =	$\frac{5}{100}$ =	$\frac{9}{12}$ =
$\frac{4}{10}$ =	$\frac{4}{12}$ =	$\frac{2}{10}$ =	$\frac{3}{6}$ =	$\frac{25}{100}$ =	$\frac{3}{12}$ =
$\frac{4}{16}$ =	$\frac{3}{9}$ =	$\frac{6}{9}$ =	$\frac{4}{8}$ =	$\frac{2}{12}$ =	$\frac{6}{12}$ =
$\frac{8}{16}$ =	$\frac{2}{6}$ =	$\frac{8}{12}$ =	$\frac{6}{8}$ =	$\frac{5}{10}$ =	$\frac{75}{100}$ =

Mental Math

a.	**b.**	**c.**	**d.**
e.	**f.**	**g.**	**h.**

Problem Solving

Understand

What information am I given?

What am I asked to find or do?

Plan

How can I use the information I am given?

Which strategy should I try?

Solve

Did I follow the plan?

Did I show my work?

Did I write the answer?

Check

Did I use the correct information?

Did I do what was asked?

Is my answer reasonable?

Name ______________________ Time __________

Power Up I

Use with Lesson 55

Facts Write each improper fraction as a mixed number. Reduce fractions.

$\frac{5}{4}$ =	$\frac{6}{4}$ =	$\frac{15}{10}$ =	$\frac{8}{3}$ =	$\frac{15}{12}$ =
$\frac{12}{8}$ =	$\frac{10}{8}$ =	$\frac{3}{2}$ =	$\frac{15}{6}$ =	$\frac{10}{4}$ =
$\frac{8}{6}$ =	$\frac{25}{10}$ =	$\frac{9}{6}$ =	$\frac{10}{6}$ =	$\frac{15}{8}$ =
$\frac{12}{10}$ =	$\frac{10}{3}$ =	$\frac{18}{12}$ =	$\frac{5}{2}$ =	$\frac{4}{3}$ =

Mental Math

a.	**b.**	**c.**	**d.**
e.	**f.**	**g.**	**h.**

Problem Solving

Understand

What information am I given?

What am I asked to find or do?

Plan

How can I use the information I am given?

Which strategy should I try?

Solve

Did I follow the plan?

Did I show my work?

Did I write the answer?

Check

Did I use the correct information?

Did I do what was asked?

Is my answer reasonable?

Name ______________________ Time __________

Power Up G

Use with Lesson 56

Facts Reduce each fraction to lowest terms.

$\frac{2}{8}$ =	$\frac{4}{6}$ =	$\frac{6}{10}$ =	$\frac{2}{4}$ =	$\frac{5}{100}$ =	$\frac{9}{12}$ =
$\frac{4}{10}$ =	$\frac{4}{12}$ =	$\frac{2}{10}$ =	$\frac{3}{6}$ =	$\frac{25}{100}$ =	$\frac{3}{12}$ =
$\frac{4}{16}$ =	$\frac{3}{9}$ =	$\frac{6}{9}$ =	$\frac{4}{8}$ =	$\frac{2}{12}$ =	$\frac{6}{12}$ =
$\frac{8}{16}$ =	$\frac{2}{6}$ =	$\frac{8}{12}$ =	$\frac{6}{8}$ =	$\frac{5}{10}$ =	$\frac{75}{100}$ =

Mental Math

a.	**b.**	**c.**	**d.**
e.	**f.**	**g.**	**h.**

Problem Solving

Understand
What information am I given?
What am I asked to find or do?

Plan
How can I use the information I am given?
Which strategy should I try?

Solve
Did I follow the plan?
Did I show my work?
Did I write the answer?

Check
Did I use the correct information?
Did I do what was asked?
Is my answer reasonable?

Name ____________________ Time __________

Power Up H

Use with Lesson 57

Facts Multiply or divide as indicated.

4×9	$4\overline{)16}$	6×8	$3\overline{)12}$	5×7	$4\overline{)32}$	3×9	$9\overline{)81}$	6×2	$8\overline{)64}$
9×7	$8\overline{)40}$	2×4	$6\overline{)42}$	5×5	$7\overline{)14}$	7×7	$8\overline{)8}$	3×3	$6\overline{)0}$
7×3	$2\overline{)10}$	10×10	$3\overline{)24}$	4×5	$9\overline{)54}$	9×1	$3\overline{)6}$	7×4	$7\overline{)56}$
6×6	$2\overline{)18}$	3×5	$5\overline{)30}$	2×2	$6\overline{)18}$	9×5	$6\overline{)24}$	2×8	$9\overline{)72}$

Mental Math

a.	**b.**	**c.**	**d.**
e.	**f.**	**g.**	**h.**

Problem Solving

Understand
What information am I given?
What am I asked to find or do?

Plan
How can I use the information I am given?
Which strategy should I try?

Solve
Did I follow the plan?
Did I show my work?
Did I write the answer?

Check
Did I use the correct information?
Did I do what was asked?
Is my answer reasonable?

Name ______________________ Time ____________

Power Up I

Use with Lesson 58

Facts Write each improper fraction as a mixed number. Reduce fractions.

$\frac{5}{4} =$	$\frac{6}{4} =$	$\frac{15}{10} =$	$\frac{8}{3} =$	$\frac{15}{12} =$
$\frac{12}{8} =$	$\frac{10}{8} =$	$\frac{3}{2} =$	$\frac{15}{6} =$	$\frac{10}{4} =$
$\frac{8}{6} =$	$\frac{25}{10} =$	$\frac{9}{6} =$	$\frac{10}{6} =$	$\frac{15}{8} =$
$\frac{12}{10} =$	$\frac{10}{3} =$	$\frac{18}{12} =$	$\frac{5}{2} =$	$\frac{4}{3} =$

Mental Math

a.	**b.**	**c.**	**d.**
e.	**f.**	**g.**	**h.**

Problem Solving

Understand

What information am I given?

What am I asked to find or do?

Plan

How can I use the information I am given?

Which strategy should I try?

Solve

Did I follow the plan?

Did I show my work?

Did I write the answer?

Check

Did I use the correct information?

Did I do what was asked?

Is my answer reasonable?

Name ______________________ Time __________

Power Up G

Use with Lesson 59

Facts Reduce each fraction to lowest terms.

$\frac{2}{8} =$	$\frac{4}{6} =$	$\frac{6}{10} =$	$\frac{2}{4} =$	$\frac{5}{100} =$	$\frac{9}{12} =$
$\frac{4}{10} =$	$\frac{4}{12} =$	$\frac{2}{10} =$	$\frac{3}{6} =$	$\frac{25}{100} =$	$\frac{3}{12} =$
$\frac{4}{16} =$	$\frac{3}{9} =$	$\frac{6}{9} =$	$\frac{4}{8} =$	$\frac{2}{12} =$	$\frac{6}{12} =$
$\frac{8}{16} =$	$\frac{2}{6} =$	$\frac{8}{12} =$	$\frac{6}{8} =$	$\frac{5}{10} =$	$\frac{75}{100} =$

Mental Math

a.	**b.**	**c.**	**d.**
e.	**f.**	**g.**	**h.**

Problem Solving

Understand

What information am I given?

What am I asked to find or do?

Plan

How can I use the information I am given?

Which strategy should I try?

Solve

Did I follow the plan?

Did I show my work?

Did I write the answer?

Check

Did I use the correct information?

Did I do what was asked?

Is my answer reasonable?

Name ______________________ Time ____________

Power Up I

Use with Lesson 60

Facts Write each improper fraction as a mixed number. Reduce fractions.

$\frac{5}{4} =$	$\frac{6}{4} =$	$\frac{15}{10} =$	$\frac{8}{3} =$	$\frac{15}{12} =$
$\frac{12}{8} =$	$\frac{10}{8} =$	$\frac{3}{2} =$	$\frac{15}{6} =$	$\frac{10}{4} =$
$\frac{8}{6} =$	$\frac{25}{10} =$	$\frac{9}{6} =$	$\frac{10}{6} =$	$\frac{15}{8} =$
$\frac{12}{10} =$	$\frac{10}{3} =$	$\frac{18}{12} =$	$\frac{5}{2} =$	$\frac{4}{3} =$

Mental Math

a.	**b.**	**c.**	**d.**
e.	**f.**	**g.**	**h.**

Problem Solving

Understand

What information am I given?

What am I asked to find or do?

Plan

How can I use the information I am given?

Which strategy should I try?

Solve

Did I follow the plan?

Did I show my work?

Did I write the answer?

Check

Did I use the correct information?

Did I do what was asked?

Is my answer reasonable?

Name ______________________ Time __________

Power Up H

Use with Lesson 61

Facts

Multiply or divide as indicated.

4×9	$4\overline{)16}$	6×8	$3\overline{)12}$	5×7	$4\overline{)32}$	3×9	$9\overline{)81}$	6×2	$8\overline{)64}$
9×7	$8\overline{)40}$	2×4	$6\overline{)42}$	5×5	$7\overline{)14}$	7×7	$8\overline{)8}$	3×3	$6\overline{)0}$
7×3	$2\overline{)10}$	10×10	$3\overline{)24}$	4×5	$9\overline{)54}$	9×1	$3\overline{)6}$	7×4	$7\overline{)56}$
6×6	$2\overline{)18}$	3×5	$5\overline{)30}$	2×2	$6\overline{)18}$	9×5	$6\overline{)24}$	2×8	$9\overline{)72}$

Mental Math

a.	**b.**	**c.**	**d.**
e.	**f.**	**g.**	**h.**

Problem Solving

Understand

What information am I given?

What am I asked to find or do?

Plan

How can I use the information I am given?

Which strategy should I try?

Solve

Did I follow the plan?

Did I show my work?

Did I write the answer?

Check

Did I use the correct information?

Did I do what was asked?

Is my answer reasonable?

Name ______________________ Time __________

Power Up G

Use with Lesson 62

Facts

Reduce each fraction to lowest terms.

$\frac{2}{8}$ =	$\frac{4}{6}$ =	$\frac{6}{10}$ =	$\frac{2}{4}$ =	$\frac{5}{100}$ =	$\frac{9}{12}$ =
$\frac{4}{10}$ =	$\frac{4}{12}$ =	$\frac{2}{10}$ =	$\frac{3}{6}$ =	$\frac{25}{100}$ =	$\frac{3}{12}$ =
$\frac{4}{16}$ =	$\frac{3}{9}$ =	$\frac{6}{9}$ =	$\frac{4}{8}$ =	$\frac{2}{12}$ =	$\frac{6}{12}$ =
$\frac{8}{16}$ =	$\frac{2}{6}$ =	$\frac{8}{12}$ =	$\frac{6}{8}$ =	$\frac{5}{10}$ =	$\frac{75}{100}$ =

Mental Math

a.	**b.**	**c.**	**d.**
e.	**f.**	**g.**	**h.**

Problem Solving

Understand

What information am I given?

What am I asked to find or do?

Plan

How can I use the information I am given?

Which strategy should I try?

Solve

Did I follow the plan?

Did I show my work?

Did I write the answer?

Check

Did I use the correct information?

Did I do what was asked?

Is my answer reasonable?

Name ______________________ Time ____________

Power Up D

Use with Lesson 63

Facts Multiply.

7 × 7	4 × 6	8 × 1	2 × 2	0 × 5	6 × 3	8 × 9	5 × 8	6 × 2	10 × 10
9 × 4	2 × 5	9 × 6	7 × 3	5 × 5	7 × 2	6 × 8	3 × 5	9 × 9	5 × 4
3 × 4	6 × 5	8 × 2	4 × 4	6 × 7	8 × 8	2 × 3	7 × 4	5 × 9	3 × 8
3 × 9	7 × 8	2 × 4	5 × 7	3 × 3	9 × 7	4 × 8	0 × 0	9 × 2	6 × 6

Mental Math

a.	**b.**	**c.**	**d.**
e.	**f.**	**g.**	**h.**

Problem Solving

Understand
What information am I given?
What am I asked to find or do?

Plan
How can I use the information I am given?
Which strategy should I try?

Solve
Did I follow the plan?
Did I show my work?
Did I write the answer?

Check
Did I use the correct information?
Did I do what was asked?
Is my answer reasonable?

Name ____________________ Time __________

Power Up J

Use with Lesson 64

Facts

Write each mixed number as an improper fraction.

$2\frac{1}{2} =$	$2\frac{2}{5} =$	$1\frac{3}{4} =$	$2\frac{3}{4} =$	$2\frac{1}{8} =$
$1\frac{2}{3} =$	$3\frac{1}{2} =$	$1\frac{5}{6} =$	$2\frac{1}{4} =$	$1\frac{1}{8} =$
$5\frac{1}{2} =$	$1\frac{3}{8} =$	$5\frac{1}{3} =$	$3\frac{1}{4} =$	$4\frac{1}{2} =$
$1\frac{7}{8} =$	$2\frac{2}{3} =$	$1\frac{5}{8} =$	$3\frac{3}{4} =$	$7\frac{1}{2} =$

Mental Math

a.	**b.**	**c.**	**d.**
e.	**f.**	**g.**	**h.**

Problem Solving

Understand

What information am I given?

What am I asked to find or do?

Plan

How can I use the information I am given?

Which strategy should I try?

Solve

Did I follow the plan?

Did I show my work?

Did I write the answer?

Check

Did I use the correct information?

Did I do what was asked?

Is my answer reasonable?

Name ______________________ Time ____________

Power Up H

Use with Lesson 65

Facts Multiply or divide as indicated.

4×9	$4\overline{)16}$	6×8	$3\overline{)12}$	5×7	$4\overline{)32}$	3×9	$9\overline{)81}$	6×2	$8\overline{)64}$
9×7	$8\overline{)40}$	2×4	$6\overline{)42}$	5×5	$7\overline{)14}$	7×7	$8\overline{)8}$	3×3	$6\overline{)0}$
7×3	$2\overline{)10}$	10×10	$3\overline{)24}$	4×5	$9\overline{)54}$	9×1	$3\overline{)6}$	7×4	$7\overline{)56}$
6×6	$2\overline{)18}$	3×5	$5\overline{)30}$	2×2	$6\overline{)18}$	9×5	$6\overline{)24}$	2×8	$9\overline{)72}$

Mental Math

a.	**b.**	**c.**	**d.**
e.	**f.**	**g.**	**h.**

Problem Solving

Understand

What information am I given?

What am I asked to find or do?

Plan

How can I use the information I am given?

Which strategy should I try?

Solve

Did I follow the plan?

Did I show my work?

Did I write the answer?

Check

Did I use the correct information?

Did I do what was asked?

Is my answer reasonable?

Name ____________________ Time ____________

Power Up J

Use with Lesson 66

Facts Write each mixed number as an improper fraction.

$2\frac{1}{2} =$	$2\frac{2}{5} =$	$1\frac{3}{4} =$	$2\frac{3}{4} =$	$2\frac{1}{8} =$
$1\frac{2}{3} =$	$3\frac{1}{2} =$	$1\frac{5}{6} =$	$2\frac{1}{4} =$	$1\frac{1}{8} =$
$5\frac{1}{2} =$	$1\frac{3}{8} =$	$5\frac{1}{3} =$	$3\frac{1}{4} =$	$4\frac{1}{2} =$
$1\frac{7}{8} =$	$2\frac{2}{3} =$	$1\frac{5}{8} =$	$3\frac{3}{4} =$	$7\frac{1}{2} =$

Mental Math

a.	b.	c.	d.
e.	f.	g.	h.

Problem Solving

Understand
What information am I given?
What am I asked to find or do?

Plan
How can I use the information I am given?
Which strategy should I try?

Solve
Did I follow the plan?
Did I show my work?
Did I write the answer?

Check
Did I use the correct information?
Did I do what was asked?
Is my answer reasonable?

Name ______________________ Time __________

Power Up J

Use with Lesson 67

Facts Write each mixed number as an improper fraction.

$2\frac{1}{2} =$	$2\frac{2}{5} =$	$1\frac{3}{4} =$	$2\frac{3}{4} =$	$2\frac{1}{8} =$
$1\frac{2}{3} =$	$3\frac{1}{2} =$	$1\frac{5}{6} =$	$2\frac{1}{4} =$	$1\frac{1}{8} =$
$5\frac{1}{2} =$	$1\frac{3}{8} =$	$5\frac{1}{3} =$	$3\frac{1}{4} =$	$4\frac{1}{2} =$
$1\frac{7}{8} =$	$2\frac{2}{3} =$	$1\frac{5}{8} =$	$3\frac{3}{4} =$	$7\frac{1}{2} =$

Mental Math

a.	**b.**	**c.**	**d.**
e.	**f.**	**g.**	**h.**

Problem Solving

Understand
What information am I given?
What am I asked to find or do?

Plan
How can I use the information I am given?
Which strategy should I try?

Solve
Did I follow the plan?
Did I show my work?
Did I write the answer?

Check
Did I use the correct information?
Did I do what was asked?
Is my answer reasonable?

Name ______________________ Time ____________

Power Up I

Use with Lesson 68

Facts Write each improper fraction as a mixed number. Reduce fractions.

$\frac{5}{4} =$	$\frac{6}{4} =$	$\frac{15}{10} =$	$\frac{8}{3} =$	$\frac{15}{12} =$
$\frac{12}{8} =$	$\frac{10}{8} =$	$\frac{3}{2} =$	$\frac{15}{6} =$	$\frac{10}{4} =$
$\frac{8}{6} =$	$\frac{25}{10} =$	$\frac{9}{6} =$	$\frac{10}{6} =$	$\frac{15}{8} =$
$\frac{12}{10} =$	$\frac{10}{3} =$	$\frac{18}{12} =$	$\frac{5}{2} =$	$\frac{4}{3} =$

Mental Math

a.	**b.**	**c.**	**d.**
e.	**f.**	**g.**	**h.**

Problem Solving

Understand
What information am I given?
What am I asked to find or do?

Plan
How can I use the information I am given?
Which strategy should I try?

Solve
Did I follow the plan?
Did I show my work?
Did I write the answer?

Check
Did I use the correct information?
Did I do what was asked?
Is my answer reasonable?

Name ________________ Time __________

Power Up J

Use with Lesson 69

Facts Write each mixed number as an improper fraction.

$2\frac{1}{2} =$	$2\frac{2}{5} =$	$1\frac{3}{4} =$	$2\frac{3}{4} =$	$2\frac{1}{8} =$
$1\frac{2}{3} =$	$3\frac{1}{2} =$	$1\frac{5}{6} =$	$2\frac{1}{4} =$	$1\frac{1}{8} =$
$5\frac{1}{2} =$	$1\frac{3}{8} =$	$5\frac{1}{3} =$	$3\frac{1}{4} =$	$4\frac{1}{2} =$
$1\frac{7}{8} =$	$2\frac{2}{3} =$	$1\frac{5}{8} =$	$3\frac{3}{4} =$	$7\frac{1}{2} =$

Mental Math

a.	**b.**	**c.**	**d.**
e.	**f.**	**g.**	**h.**

Problem Solving

Understand
What information am I given?
What am I asked to find or do?

Plan
How can I use the information I am given?
Which strategy should I try?

Solve
Did I follow the plan?
Did I show my work?
Did I write the answer?

Check
Did I use the correct information?
Did I do what was asked?
Is my answer reasonable?

Name ____________________ Time __________

Power Up G

Use with Lesson 70

Facts Reduce each fraction to lowest terms.

$\frac{2}{8} =$	$\frac{4}{6} =$	$\frac{6}{10} =$	$\frac{2}{4} =$	$\frac{5}{100} =$	$\frac{9}{12} =$
$\frac{4}{10} =$	$\frac{4}{12} =$	$\frac{2}{10} =$	$\frac{3}{6} =$	$\frac{25}{100} =$	$\frac{3}{12} =$
$\frac{4}{16} =$	$\frac{3}{9} =$	$\frac{6}{9} =$	$\frac{4}{8} =$	$\frac{2}{12} =$	$\frac{6}{12} =$
$\frac{8}{16} =$	$\frac{2}{6} =$	$\frac{8}{12} =$	$\frac{6}{8} =$	$\frac{5}{10} =$	$\frac{75}{100} =$

Mental Math

a.	**b.**	**c.**	**d.**
e.	**f.**	**g.**	**h.**

Problem Solving

Understand

What information am I given?

What am I asked to find or do?

Plan

How can I use the information I am given?

Which strategy should I try?

Solve

Did I follow the plan?

Did I show my work?

Did I write the answer?

Check

Did I use the correct information?

Did I do what was asked?

Is my answer reasonable?

Name ____________________ Time __________

Power Up D

Use with Lesson 71

Facts

Multiply.

7 × 7	4 × 6	8 × 1	2 × 2	0 × 5	6 × 3	8 × 9	5 × 8	6 × 2	10 × 10
9 × 4	2 × 5	9 × 6	7 × 3	5 × 5	7 × 2	6 × 8	3 × 5	9 × 9	5 × 4
3 × 4	6 × 5	8 × 2	4 × 4	6 × 7	8 × 8	2 × 3	7 × 4	5 × 9	3 × 8
3 × 9	7 × 8	2 × 4	5 × 7	3 × 3	9 × 7	4 × 8	0 × 0	9 × 2	6 × 6

Mental Math

a.	**b.**	**c.**	**d.**
e.	**f.**	**g.**	**h.**

Problem Solving

Understand

What information am I given?

What am I asked to find or do?

Plan

How can I use the information I am given?

Which strategy should I try?

Solve

Did I follow the plan?

Did I show my work?

Did I write the answer?

Check

Did I use the correct information?

Did I do what was asked?

Is my answer reasonable?

Name ____________________ Time __________

Power Up H

Use with Lesson 72

Facts

Multiply or divide as indicated.

4×9	$4\overline{)16}$	6×8	$3\overline{)12}$	5×7	$4\overline{)32}$	3×9	$9\overline{)81}$	6×2	$8\overline{)64}$
9×7	$8\overline{)40}$	2×4	$6\overline{)42}$	5×5	$7\overline{)14}$	7×7	$8\overline{)8}$	3×3	$6\overline{)0}$
7×3	$2\overline{)10}$	10×10	$3\overline{)24}$	4×5	$9\overline{)54}$	9×1	$3\overline{)6}$	7×4	$7\overline{)56}$
6×6	$2\overline{)18}$	3×5	$5\overline{)30}$	2×2	$6\overline{)18}$	9×5	$6\overline{)24}$	2×8	$9\overline{)72}$

Mental Math

a.	b.	c.	d.
e.	f.	g.	h.

Problem Solving

Understand
What information am I given?
What am I asked to find or do?

Plan
How can I use the information I am given?
Which strategy should I try?

Solve
Did I follow the plan?
Did I show my work?
Did I write the answer?

Check
Did I use the correct information?
Did I do what was asked?
Is my answer reasonable?

Name ______________________ Time ____________

Power Up J

Use with Lesson 73

Facts

Write each mixed number as an improper fraction.

$2\frac{1}{2}$ =	$2\frac{2}{5}$ =	$1\frac{3}{4}$ =	$2\frac{3}{4}$ =	$2\frac{1}{8}$ =
$1\frac{2}{3}$ =	$3\frac{1}{2}$ =	$1\frac{5}{6}$ =	$2\frac{1}{4}$ =	$1\frac{1}{8}$ =
$5\frac{1}{2}$ =	$1\frac{3}{8}$ =	$5\frac{1}{3}$ =	$3\frac{1}{4}$ =	$4\frac{1}{2}$ =
$1\frac{7}{8}$ =	$2\frac{2}{3}$ =	$1\frac{5}{8}$ =	$3\frac{3}{4}$ =	$7\frac{1}{2}$ =

Mental Math

a.	**b.**	**c.**	**d.**
e.	**f.**	**g.**	**h.**

Problem Solving

Understand
What information am I given?
What am I asked to find or do?

Plan
How can I use the information I am given?
Which strategy should I try?

Solve
Did I follow the plan?
Did I show my work?
Did I write the answer?

Check
Did I use the correct information?
Did I do what was asked?
Is my answer reasonable?

Name ______________________ Time ____________

Power Up I

Use with Lesson 74

Facts

Write each improper fraction as a mixed number. Reduce fractions.

$\frac{5}{4} =$	$\frac{6}{4} =$	$\frac{15}{10} =$	$\frac{8}{3} =$	$\frac{15}{12} =$
$\frac{12}{8} =$	$\frac{10}{8} =$	$\frac{3}{2} =$	$\frac{15}{6} =$	$\frac{10}{4} =$
$\frac{8}{6} =$	$\frac{25}{10} =$	$\frac{9}{6} =$	$\frac{10}{6} =$	$\frac{15}{8} =$
$\frac{12}{10} =$	$\frac{10}{3} =$	$\frac{18}{12} =$	$\frac{5}{2} =$	$\frac{4}{3} =$

Mental Math

a.	**b.**	**c.**	**d.**
e.	**f.**	**g.**	**h.**

Problem Solving

Understand

What information am I given?

What am I asked to find or do?

Plan

How can I use the information I am given?

Which strategy should I try?

Solve

Did I follow the plan?

Did I show my work?

Did I write the answer?

Check

Did I use the correct information?

Did I do what was asked?

Is my answer reasonable?

Name ____________________ Time ____________

Power Up K

Use with Lesson 75

Facts

Complete each equivalent measure.		Write a unit for each reference.
1. 1 cm = ______ mm	13. 10 cm = ______ mm	Metric Units:
2. 1 m = ______ mm	14. 2 m = ______ cm	25. The thickness of a dime: ______
3. 1 m = ______ cm	15. 5 km = ______ m	26. The width of a little finger: ______
4. 1 km = ______ m	16. 2.5 cm = ______ mm	
5. 1 in. = ______ cm	17. 1.5 m = ______ cm	27. The length of one big step: ______
6. 1 mi ≈ ______ m	18. 7.5 km = ______ m	
7. 1 ft = ______ in.	19. $\frac{1}{2}$ ft = ______ in.	U.S. Customary Units:
8. 1 yd = ______ in.	20. 2 ft = ______ in.	28. The width of two fingers: ______
9. 1 yd = ______ ft	21. 3 ft = ______ in.	29. The length of a man's shoe: ______
10. 1 mi = ______ ft	22. 2 yd = ______ ft	
11. 1 m ≈ ______ in.	23. 10 yd = ______ ft	30. The length of one big step: ______
12. 1 km ≈ ______ mi	24. 100 yd = ______ ft	

Mental Math

a.	b.	c.	d.
e.	f.	g.	h.

Problem Solving

Understand
What information am I given?
What am I asked to find or do?

Plan
How can I use the information I am given?
Which strategy should I try?

Solve
Did I follow the plan?
Did I show my work?
Did I write the answer?

Check
Did I use the correct information?
Did I do what was asked?
Is my answer reasonable?

Name ______________________ Time ____________

Power Up G

Use with Lesson 76

Facts Reduce each fraction to lowest terms.

$\frac{2}{8}$ =	$\frac{4}{6}$ =	$\frac{6}{10}$ =	$\frac{2}{4}$ =	$\frac{5}{100}$ =	$\frac{9}{12}$ =
$\frac{4}{10}$ =	$\frac{4}{12}$ =	$\frac{2}{10}$ =	$\frac{3}{6}$ =	$\frac{25}{100}$ =	$\frac{3}{12}$ =
$\frac{4}{16}$ =	$\frac{3}{9}$ =	$\frac{6}{9}$ =	$\frac{4}{8}$ =	$\frac{2}{12}$ =	$\frac{6}{12}$ =
$\frac{8}{16}$ =	$\frac{2}{6}$ =	$\frac{8}{12}$ =	$\frac{6}{8}$ =	$\frac{5}{10}$ =	$\frac{75}{100}$ =

Mental Math

a.	**b.**	**c.**	**d.**
e.	**f.**	**g.**	**h.**

Problem Solving

Understand

What information am I given?

What am I asked to find or do?

Plan

How can I use the information I am given?

Which strategy should I try?

Solve

Did I follow the plan?

Did I show my work?

Did I write the answer?

Check

Did I use the correct information?

Did I do what was asked?

Is my answer reasonable?

Name ____________ Time ________

Power Up K

Use with Lesson 77

Facts

Complete each equivalent measure.		Write a unit for each reference.
1. 1 cm = ______ mm	13. 10 cm = ______ mm	Metric Units:
2. 1 m = ______ mm	14. 2 m = ______ cm	25. The thickness of a dime: ______
3. 1 m = ______ cm	15. 5 km = ______ m	26. The width of a little finger: ______
4. 1 km = ______ m	16. 2.5 cm = ______ mm	27. The length of one big step: ______
5. 1 in. = ______ cm	17. 1.5 m = ______ cm	
6. 1 mi ≈ ______ m	18. 7.5 km = ______ m	
7. 1 ft = ______ in.	19. $\frac{1}{2}$ ft = ______ in.	U.S. Customary Units:
8. 1 yd = ______ in.	20. 2 ft = ______ in.	28. The width of two fingers: ______
9. 1 yd = ______ ft	21. 3 ft = ______ in.	29. The length of a man's shoe: ______
10. 1 mi = ______ ft	22. 2 yd = ______ ft	30. The length of one big step: ______
11. 1 m ≈ ______ in.	23. 10 yd = ______ ft	
12. 1 km ≈ ______ mi	24. 100 yd = ______ ft	

Mental Math

a.	b.	c.	d.
e.	f.	g.	h.

Problem Solving

Understand
What information am I given?
What am I asked to find or do?

Plan
How can I use the information I am given?
Which strategy should I try?

Solve
Did I follow the plan?
Did I show my work?
Did I write the answer?

Check
Did I use the correct information?
Did I do what was asked?
Is my answer reasonable?

Name _______________ Time _______

Power Up J

Use with Lesson 78

Facts

Write each mixed number as an improper fraction.

$2\frac{1}{2} =$	$2\frac{2}{5} =$	$1\frac{3}{4} =$	$2\frac{3}{4} =$	$2\frac{1}{8} =$
$1\frac{2}{3} =$	$3\frac{1}{2} =$	$1\frac{5}{6} =$	$2\frac{1}{4} =$	$1\frac{1}{8} =$
$5\frac{1}{2} =$	$1\frac{3}{8} =$	$5\frac{1}{3} =$	$3\frac{1}{4} =$	$4\frac{1}{2} =$
$1\frac{7}{8} =$	$2\frac{2}{3} =$	$1\frac{5}{8} =$	$3\frac{3}{4} =$	$7\frac{1}{2} =$

Mental Math

a.	**b.**	**c.**	**d.**
e.	**f.**	**g.**	**h.**

Problem Solving

Understand

What information am I given?

What am I asked to find or do?

Plan

How can I use the information I am given?

Which strategy should I try?

Solve

Did I follow the plan?

Did I show my work?

Did I write the answer?

Check

Did I use the correct information?

Did I do what was asked?

Is my answer reasonable?

Name ______________________ Time __________

Power Up K

Use with Lesson 79

Facts

Complete each equivalent measure.		Write a unit for each reference.
1. 1 cm = ______ mm 2. 1 m = ______ mm 3. 1 m = ______ cm 4. 1 km = ______ m	13. 10 cm = ______ mm 14. 2 m = ______ cm 15. 5 km = ______ m 16. 2.5 cm = ______ mm 17. 1.5 m = ______ cm 18. 7.5 km = ______ m	Metric Units: 25. The thickness of a dime: ______ 26. The width of a little finger: ______ 27. The length of one big step: ______
5. 1 in. = ______ cm 6. 1 mi ≈ ______ m		
7. 1 ft = ______ in. 8. 1 yd = ______ in. 9. 1 yd = ______ ft 10. 1 mi = ______ ft	19. $\frac{1}{2}$ ft = ______ in. 20. 2 ft = ______ in. 21. 3 ft = ______ in. 22. 2 yd = ______ ft 23. 10 yd = ______ ft 24. 100 yd = ______ ft	U.S. Customary Units: 28. The width of two fingers: ______ 29. The length of a man's shoe: ______ 30. The length of one big step: ______
11. 1 m ≈ ______ in. 12. 1 km ≈ ______ mi		

Mental Math

a.	b.	c.	d.
e.	f.	g.	h.

Problem Solving

Understand
What information am I given?
What am I asked to find or do?

Plan
How can I use the information I am given?
Which strategy should I try?

Solve
Did I follow the plan?
Did I show my work?
Did I write the answer?

Check
Did I use the correct information?
Did I do what was asked?
Is my answer reasonable?

Name ____________________ Time __________

Power Up I

Use with Lesson 80

Facts

Write each improper fraction as a mixed number. Reduce fractions.

$\frac{5}{4} =$	$\frac{6}{4} =$	$\frac{15}{10} =$	$\frac{8}{3} =$	$\frac{15}{12} =$
$\frac{12}{8} =$	$\frac{10}{8} =$	$\frac{3}{2} =$	$\frac{15}{6} =$	$\frac{10}{4} =$
$\frac{8}{6} =$	$\frac{25}{10} =$	$\frac{9}{6} =$	$\frac{10}{6} =$	$\frac{15}{8} =$
$\frac{12}{10} =$	$\frac{10}{3} =$	$\frac{18}{12} =$	$\frac{5}{2} =$	$\frac{4}{3} =$

Mental Math

a.	**b.**	**c.**	**d.**
e.	**f.**	**g.**	**h.**

Problem Solving

Understand

What information am I given?

What am I asked to find or do?

Plan

How can I use the information I am given?

Which strategy should I try?

Solve

Did I follow the plan?

Did I show my work?

Did I write the answer?

Check

Did I use the correct information?

Did I do what was asked?

Is my answer reasonable?

Name ____________________ Time ____________

Power Up K

Use with Lesson 81

Facts

Complete each equivalent measure.		Write a unit for each reference.
1. 1 cm = ______ mm	13. 10 cm = ______ mm	Metric Units:
2. 1 m = ______ mm	14. 2 m = ______ cm	25. The thickness of a dime: ______
3. 1 m = ______ cm	15. 5 km = ______ m	26. The width of a little finger: ______
4. 1 km = ______ m	16. 2.5 cm = ______ mm	
5. 1 in. = ______ cm	17. 1.5 m = ______ cm	27. The length of one big step: ______
6. 1 mi ≈ ______ m	18. 7.5 km = ______ m	
		U.S. Customary Units:
7. 1 ft = ______ in.	19. $\frac{1}{2}$ ft = ______ in.	28. The width of two fingers: ______
8. 1 yd = ______ in.	20. 2 ft = ______ in.	
9. 1 yd = ______ ft	21. 3 ft = ______ in.	29. The length of a man's shoe: ______
10. 1 mi = ______ ft	22. 2 yd = ______ ft	
11. 1 m ≈ ______ in.	23. 10 yd = ______ ft	30. The length of one big step: ______
12. 1 km ≈ ______ mi	24. 100 yd = ______ ft	

Mental Math

a.	b.	c.	d.
e.	f.	g.	h.

Problem Solving

Understand
What information am I given?
What am I asked to find or do?

Plan
How can I use the information I am given?
Which strategy should I try?

Solve
Did I follow the plan?
Did I show my work?
Did I write the answer?

Check
Did I use the correct information?
Did I do what was asked?
Is my answer reasonable?

Name ______________________ Time __________

Power Up L

Use with Lesson 82

Facts

Write the abbreviation.	Complete each equivalence.	Complete each conversion.
Metric Units:	Metric Units:	14. 2 liters = ______ milliliters
1. liter ______	7. 1 liter = ______ milliliters	15. 2 liters ≈ ______ quarts
2. milliliter ______	U.S. Customary Units:	16. 3.78 liters = ______ milliliters
U.S. Customary Units:	8. 1 cup = ______ ounces	17. 0.5 liter = ______ milliliters
3. ounces ______	9. 1 pint = ______ ounces	18. $\frac{1}{2}$ gallon = ______ quarts
4. pint ______	10. 1 pint = ______ cups	19. 2 gallons = ______ quarts
5. quart ______	11. 1 quart = ______ pints	20. 2 half gallons = ______ gallon
6. gallon ______	12. 1 gallon = ______ quarts	21. 8 cups = ______ quarts
	Between Systems:	22–23. A two-liter bottle is a little more than ______ quarts or ______ gallon.
	13. 1 liter ≈ ______ quart	

Mental Math

a.	b.	c.	d.
e.	f.	g.	h.

Problem Solving

Understand

What information am I given?

What am I asked to find or do?

Plan

How can I use the information I am given?

Which strategy should I try?

Solve

Did I follow the plan?

Did I show my work?

Did I write the answer?

Check

Did I use the correct information?

Did I do what was asked?

Is my answer reasonable?

Name ______________________ Time ____________

Power Up I

Use with Lesson 83

Facts Write each improper fraction as a mixed number. Reduce fractions.

$\frac{5}{4} =$	$\frac{6}{4} =$	$\frac{15}{10} =$	$\frac{8}{3} =$	$\frac{15}{12} =$
$\frac{12}{8} =$	$\frac{10}{8} =$	$\frac{3}{2} =$	$\frac{15}{6} =$	$\frac{10}{4} =$
$\frac{8}{6} =$	$\frac{25}{10} =$	$\frac{9}{6} =$	$\frac{10}{6} =$	$\frac{15}{8} =$
$\frac{12}{10} =$	$\frac{10}{3} =$	$\frac{18}{12} =$	$\frac{5}{2} =$	$\frac{4}{3} =$

Mental Math

a.	**b.**	**c.**	**d.**
e.	**f.**	**g.**	**h.**

Problem Solving

Understand

What information am I given?

What am I asked to find or do?

Plan

How can I use the information I am given?

Which strategy should I try?

Solve

Did I follow the plan?

Did I show my work?

Did I write the answer?

Check

Did I use the correct information?

Did I do what was asked?

Is my answer reasonable?

Name ______________________ Time __________

Power Up K

Use with Lesson 84

Facts

Complete each equivalent measure.		Write a unit for each reference.
		Metric Units:
1. 1 cm = ______ mm	13. 10 cm = ______ mm	25. The thickness of a dime: ______
2. 1 m = ______ mm	14. 2 m = ______ cm	
3. 1 m = ______ cm	15. 5 km = ______ m	26. The width of a little finger: ______
4. 1 km = ______ m	16. 2.5 cm = ______ mm	
5. 1 in. = ______ cm	17. 1.5 m = ______ cm	27. The length of one big step: ______
6. 1 mi ≈ ______ m	18. 7.5 km = ______ m	
		U.S. Customary Units:
7. 1 ft = ______ in.	19. $\frac{1}{2}$ ft = ______ in.	28. The width of two fingers: ______
8. 1 yd = ______ in.	20. 2 ft = ______ in.	
9. 1 yd = ______ ft	21. 3 ft = ______ in.	29. The length of a man's shoe: ______
10. 1 mi = ______ ft	22. 2 yd = ______ ft	
11. 1 m ≈ ______ in.	23. 10 yd = ______ ft	30. The length of one big step: ______
12. 1 km ≈ ______ mi	24. 100 yd = ______ ft	

Mental Math

a.	b.	c.	d.
e.	f.	g.	h.

Problem Solving

Understand
What information am I given?
What am I asked to find or do?

Plan
How can I use the information I am given?
Which strategy should I try?

Solve
Did I follow the plan?
Did I show my work?
Did I write the answer?

Check
Did I use the correct information?
Did I do what was asked?
Is my answer reasonable?

Name ________________ Time ________

Power Up L

Use with Lesson 85

Facts

Write the abbreviation.	Complete each equivalence.	Complete each conversion.
Metric Units:	Metric Units:	14. 2 liters = ______ milliliters
1. liter ______	7. 1 liter = ______ milliliters	15. 2 liters ≈ ______ quarts
2. milliliter ______	U.S. Customary Units:	16. 3.78 liters = ______ milliliters
U.S. Customary Units:	8. 1 cup = ______ ounces	17. 0.5 liter = ______ milliliters
3. ounces ______	9. 1 pint = ______ ounces	18. $\frac{1}{2}$ gallon = ______ quarts
4. pint ______	10. 1 pint = ______ cups	19. 2 gallons = ______ quarts
5. quart ______	11. 1 quart = ______ pints	20. 2 half gallons = ______ gallon
6. gallon ______	12. 1 gallon = ______ quarts	21. 8 cups = ______ quarts
	Between Systems:	22–23. A two-liter bottle is a little more than ______ quarts or ______ gallon.
	13. 1 liter ≈ ______ quart	

Mental Math

a.	b.	c.	d.
e.	f.	g.	h.

Problem Solving

Understand
What information am I given?
What am I asked to find or do?

Plan
How can I use the information I am given?
Which strategy should I try?

Solve
Did I follow the plan?
Did I show my work?
Did I write the answer?

Check
Did I use the correct information?
Did I do what was asked?
Is my answer reasonable?

Name ______________________ Time ____________

Power Up G

Use with Lesson 86

Facts

Reduce each fraction to lowest terms.

$\frac{2}{8} =$	$\frac{4}{6} =$	$\frac{6}{10} =$	$\frac{2}{4} =$	$\frac{5}{100} =$	$\frac{9}{12} =$
$\frac{4}{10} =$	$\frac{4}{12} =$	$\frac{2}{10} =$	$\frac{3}{6} =$	$\frac{25}{100} =$	$\frac{3}{12} =$
$\frac{4}{16} =$	$\frac{3}{9} =$	$\frac{6}{9} =$	$\frac{4}{8} =$	$\frac{2}{12} =$	$\frac{6}{12} =$
$\frac{8}{16} =$	$\frac{2}{6} =$	$\frac{8}{12} =$	$\frac{6}{8} =$	$\frac{5}{10} =$	$\frac{75}{100} =$

Mental Math

a.	**b.**	**c.**	**d.**
e.	**f.**	**g.**	**h.**

Problem Solving

Understand

What information am I given?

What am I asked to find or do?

Plan

How can I use the information I am given?

Which strategy should I try?

Solve

Did I follow the plan?

Did I show my work?

Did I write the answer?

Check

Did I use the correct information?

Did I do what was asked?

Is my answer reasonable?

Name ____________________ Time __________

Power Up D

Use with Lesson 87

Facts Multiply.

7×7	4×6	8×1	2×2	0×5	6×3	8×9	5×8	6×2	10×10
9×4	2×5	9×6	7×3	5×5	7×2	6×8	3×5	9×9	5×4
3×4	6×5	8×2	4×4	6×7	8×8	2×3	7×4	5×9	3×8
3×9	7×8	2×4	5×7	3×3	9×7	4×8	0×0	9×2	6×6

Mental Math

a.	**b.**	**c.**	**d.**
e.	**f.**	**g.**	**h.**

Problem Solving

Understand
What information am I given?
What am I asked to find or do?

Plan
How can I use the information I am given?
Which strategy should I try?

Solve
Did I follow the plan?
Did I show my work?
Did I write the answer?

Check
Did I use the correct information?
Did I do what was asked?
Is my answer reasonable?

Name ____________________ Time __________

Power Up L

Use with Lesson 88

Facts

Write the abbreviation.	Complete each equivalence.	Complete each conversion.
Metric Units:	Metric Units:	14. 2 liters = ______ milliliters
1. liter ______	7. 1 liter = ______ milliliters	15. 2 liters ≈ ______ quarts
2. milliliter ______	U.S. Customary Units:	16. 3.78 liters = ______ milliliters
U.S. Customary Units:	8. 1 cup = ______ ounces	17. 0.5 liter = ______ milliliters
3. ounces ______	9. 1 pint = ______ ounces	18. $\frac{1}{2}$ gallon = ______ quarts
4. pint ______	10. 1 pint = ______ cups	19. 2 gallons = ______ quarts
5. quart ______	11. 1 quart = ______ pints	20. 2 half gallons = ______ gallon
6. gallon ______	12. 1 gallon = ______ quarts	21. 8 cups = ______ quarts
	Between Systems:	22–23. A two-liter bottle is a little more than ______ quarts or ______ gallon.
	13. 1 liter ≈ ______ quart	

Mental Math

a.	b.	c.	d.
e.	f.	g.	h.

Problem Solving

Understand
What information am I given?
What am I asked to find or do?

Plan
How can I use the information I am given?
Which strategy should I try?

Solve
Did I follow the plan?
Did I show my work?
Did I write the answer?

Check
Did I use the correct information?
Did I do what was asked?
Is my answer reasonable?

Name ______________________ Time ____________

Power Up K

Use with Lesson 89

Facts

Complete each equivalent measure.		Write a unit for each reference.
		Metric Units:
1. 1 cm = ______ mm	13. 10 cm = ______ mm	25. The thickness of a dime: ______
2. 1 m = ______ mm	14. 2 m = ______ cm	
3. 1 m = ______ cm	15. 5 km = ______ m	26. The width of a little finger: ______
4. 1 km = ______ m	16. 2.5 cm = ______ mm	
5. 1 in. = ______ cm	17. 1.5 m = ______ cm	27. The length of one big step: ______
6. 1 mi ≈ ______ m	18. 7.5 km = ______ m	
		U.S. Customary Units:
7. 1 ft = ______ in.	19. $\frac{1}{2}$ ft = ______ in.	28. The width of two fingers: ______
8. 1 yd = ______ in.	20. 2 ft = ______ in.	
9. 1 yd = ______ ft	21. 3 ft = ______ in.	29. The length of a man's shoe: ______
10. 1 mi = ______ ft	22. 2 yd = ______ ft	
11. 1 m ≈ ______ in.	23. 10 yd = ______ ft	30. The length of one big step: ______
12. 1 km ≈ ______ mi	24. 100 yd = ______ ft	

Mental Math

a.	b.	c.	d.
e.	f.	g.	h.

Problem Solving

Understand
What information am I given?
What am I asked to find or do?

Plan
How can I use the information I am given?
Which strategy should I try?

Solve
Did I follow the plan?
Did I show my work?
Did I write the answer?

Check
Did I use the correct information?
Did I do what was asked?
Is my answer reasonable?

Name ______________________ Time __________

Power Up J

Use with Lesson 90

Facts Write each mixed number as an improper fraction.

$2\frac{1}{2} =$	$2\frac{2}{5} =$	$1\frac{3}{4} =$	$2\frac{3}{4} =$	$2\frac{1}{8} =$
$1\frac{2}{3} =$	$3\frac{1}{2} =$	$1\frac{5}{6} =$	$2\frac{1}{4} =$	$1\frac{1}{8} =$
$5\frac{1}{2} =$	$1\frac{3}{8} =$	$5\frac{1}{3} =$	$3\frac{1}{4} =$	$4\frac{1}{2} =$
$1\frac{7}{8} =$	$2\frac{2}{3} =$	$1\frac{5}{8} =$	$3\frac{3}{4} =$	$7\frac{1}{2} =$

Mental Math

a.	**b.**	**c.**	**d.**
e.	**f.**	**g.**	**h.**

Problem Solving

Understand

What information am I given?

What am I asked to find or do?

Plan

How can I use the information I am given?

Which strategy should I try?

Solve

Did I follow the plan?

Did I show my work?

Did I write the answer?

Check

Did I use the correct information?

Did I do what was asked?

Is my answer reasonable?

Name ______________________ Time __________

Power Up H

Use with Lesson 91

Facts

Multiply or divide as indicated.

4×9	$4\overline{)16}$	6×8	$3\overline{)12}$	5×7	$4\overline{)32}$	3×9	$9\overline{)81}$	6×2	$8\overline{)64}$
9×7	$8\overline{)40}$	2×4	$6\overline{)42}$	5×5	$7\overline{)14}$	7×7	$8\overline{)8}$	3×3	$6\overline{)0}$
7×3	$2\overline{)10}$	10×10	$3\overline{)24}$	4×5	$9\overline{)54}$	9×1	$3\overline{)6}$	7×4	$7\overline{)56}$
6×6	$2\overline{)18}$	3×5	$5\overline{)30}$	2×2	$6\overline{)18}$	9×5	$6\overline{)24}$	2×8	$9\overline{)72}$

Mental Math

a.	**b.**	**c.**	**d.**
e.	**f.**	**g.**	**h.**

Problem Solving

Understand
What information am I given?
What am I asked to find or do?

Plan
How can I use the information I am given?
Which strategy should I try?

Solve
Did I follow the plan?
Did I show my work?
Did I write the answer?

Check
Did I use the correct information?
Did I do what was asked?
Is my answer reasonable?

Name ______________________ Time ____________

Power Up L

Use with Lesson 92

Facts

Write the abbreviation.	Complete each equivalence.	Complete each conversion.
Metric Units:	Metric Units:	14. 2 liters = ______ milliliters
1. liter ______	7. 1 liter = ______ milliliters	15. 2 liters ≈ ______ quarts
2. milliliter ______	U.S. Customary Units:	16. 3.78 liters = ______ milliliters
U.S. Customary Units:	8. 1 cup = ______ ounces	17. 0.5 liter = ______ milliliters
3. ounces ______	9. 1 pint = ______ ounces	18. $\frac{1}{2}$ gallon = ______ quarts
4. pint ______	10. 1 pint = ______ cups	19. 2 gallons = ______ quarts
5. quart ______	11. 1 quart = ______ pints	20. 2 half gallons = ______ gallon
6. gallon ______	12. 1 gallon = ______ quarts	21. 8 cups = ______ quarts
	Between Systems:	22–23. A two-liter bottle is a little more than ______ quarts or ______ gallon.
	13. 1 liter ≈ ______ quart	

Mental Math

a.	b.	c.	d.
e.	f.	g.	h.

Problem Solving

Understand
What information am I given?
What am I asked to find or do?

Plan
How can I use the information I am given?
Which strategy should I try?

Solve
Did I follow the plan?
Did I show my work?
Did I write the answer?

Check
Did I use the correct information?
Did I do what was asked?
Is my answer reasonable?

Name ____________________ Time ________

Power Up I

Use with Lesson 93

Facts Write each improper fraction as a mixed number. Reduce fractions.

$\frac{5}{4} =$	$\frac{6}{4} =$	$\frac{15}{10} =$	$\frac{8}{3} =$	$\frac{15}{12} =$
$\frac{12}{8} =$	$\frac{10}{8} =$	$\frac{3}{2} =$	$\frac{15}{6} =$	$\frac{10}{4} =$
$\frac{8}{6} =$	$\frac{25}{10} =$	$\frac{9}{6} =$	$\frac{10}{6} =$	$\frac{15}{8} =$
$\frac{12}{10} =$	$\frac{10}{3} =$	$\frac{18}{12} =$	$\frac{5}{2} =$	$\frac{4}{3} =$

Mental Math

a.	**b.**	**c.**	**d.**
e.	**f.**	**g.**	**h.**

Problem Solving

Understand
What information am I given?
What am I asked to find or do?

Plan
How can I use the information I am given?
Which strategy should I try?

Solve
Did I follow the plan?
Did I show my work?
Did I write the answer?

Check
Did I use the correct information?
Did I do what was asked?
Is my answer reasonable?

Name ____________________ Time __________

Power Up K

Use with Lesson 94

Facts

Complete each equivalent measure.		Write a unit for each reference.
1. 1 cm = ______ mm	13. 10 cm = ______ mm	Metric Units:
2. 1 m = ______ mm	14. 2 m = ______ cm	25. The thickness of a dime: ______
3. 1 m = ______ cm	15. 5 km = ______ m	26. The width of a little finger: ______
4. 1 km = ______ m	16. 2.5 cm = ______ mm	
5. 1 in. = ______ cm	17. 1.5 m = ______ cm	27. The length of one big step: ______
6. 1 mi ≈ ______ m	18. 7.5 km = ______ m	
7. 1 ft = ______ in.	19. $\frac{1}{2}$ ft = ______ in.	U.S. Customary Units:
8. 1 yd = ______ in.	20. 2 ft = ______ in.	28. The width of two fingers: ______
9. 1 yd = ______ ft	21. 3 ft = ______ in.	29. The length of a man's shoe: ______
10. 1 mi = ______ ft	22. 2 yd = ______ ft	
11. 1 m ≈ ______ in.	23. 10 yd = ______ ft	30. The length of one big step: ______
12. 1 km ≈ ______ mi	24. 100 yd = ______ ft	

Mental Math

a.	b.	c.	d.
e.	f.	g.	h.

Problem Solving

Understand
What information am I given?
What am I asked to find or do?

Plan
How can I use the information I am given?
Which strategy should I try?

Solve
Did I follow the plan?
Did I show my work?
Did I write the answer?

Check
Did I use the correct information?
Did I do what was asked?
Is my answer reasonable?

Name ______________________ Time ____________

Power Up G

Use with Lesson 95

Facts Reduce each fraction to lowest terms.

$\frac{2}{8} =$	$\frac{4}{6} =$	$\frac{6}{10} =$	$\frac{2}{4} =$	$\frac{5}{100} =$	$\frac{9}{12} =$
$\frac{4}{10} =$	$\frac{4}{12} =$	$\frac{2}{10} =$	$\frac{3}{6} =$	$\frac{25}{100} =$	$\frac{3}{12} =$
$\frac{4}{16} =$	$\frac{3}{9} =$	$\frac{6}{9} =$	$\frac{4}{8} =$	$\frac{2}{12} =$	$\frac{6}{12} =$
$\frac{8}{16} =$	$\frac{2}{6} =$	$\frac{8}{12} =$	$\frac{6}{8} =$	$\frac{5}{10} =$	$\frac{75}{100} =$

Mental Math

a.	**b.**	**c.**	**d.**
e.	**f.**	**g.**	**h.**

Problem Solving

Understand

What information am I given?

What am I asked to find or do?

Plan

How can I use the information I am given?

Which strategy should I try?

Solve

Did I follow the plan?

Did I show my work?

Did I write the answer?

Check

Did I use the correct information?

Did I do what was asked?

Is my answer reasonable?

Name ________________________ Time __________

Power Up D

Use with Lesson 96

Facts

Multiply.

7×7	4×6	8×1	2×2	0×5	6×3	8×9	5×8	6×2	10×10
9×4	2×5	9×6	7×3	5×5	7×2	6×8	3×5	9×9	5×4
3×4	6×5	8×2	4×4	6×7	8×8	2×3	7×4	5×9	3×8
3×9	7×8	2×4	5×7	3×3	9×7	4×8	0×0	9×2	6×6

Mental Math

a.	**b.**	**c.**	**d.**
e.	**f.**	**g.**	**h.**

Problem Solving

Understand
What information am I given?
What am I asked to find or do?

Plan
How can I use the information I am given?
Which strategy should I try?

Solve
Did I follow the plan?
Did I show my work?
Did I write the answer?

Check
Did I use the correct information?
Did I do what was asked?
Is my answer reasonable?

Name ______________________ Time ____________

Power Up L

Use with Lesson 97

Facts

Write the abbreviation.	Complete each equivalence.	Complete each conversion.
Metric Units:	Metric Units:	14. 2 liters = ______ milliliters
1. liter ______	7. 1 liter = ______ milliliters	15. 2 liters ≈ ______ quarts
2. milliliter ______	U.S. Customary Units:	16. 3.78 liters = ______ milliliters
U.S. Customary Units:	8. 1 cup = ______ ounces	17. 0.5 liter = ______ milliliters
3. ounces ______	9. 1 pint = ______ ounces	18. $\frac{1}{2}$ gallon = ______ quarts
4. pint ______	10. 1 pint = ______ cups	19. 2 gallons = ______ quarts
5. quart ______	11. 1 quart = ______ pints	20. 2 half gallons = ______ gallon
6. gallon ______	12. 1 gallon = ______ quarts	21. 8 cups = ______ quarts
	Between Systems:	22–23. A two-liter bottle is a little more than ______ quarts or ______ gallon.
	13. 1 liter ≈ ______ quart	

Mental Math

a.	b.	c.	d.
e.	f.	g.	h.

Problem Solving

Understand
What information am I given?
What am I asked to find or do?

Plan
How can I use the information I am given?
Which strategy should I try?

Solve
Did I follow the plan?
Did I show my work?
Did I write the answer?

Check
Did I use the correct information?
Did I do what was asked?
Is my answer reasonable?

Name ______________________ Time __________

Power Up J

Use with Lesson 98

Facts Write each mixed number as an improper fraction.

$2\frac{1}{2} =$	$2\frac{2}{5} =$	$1\frac{3}{4} =$	$2\frac{3}{4} =$	$2\frac{1}{8} =$
$1\frac{2}{3} =$	$3\frac{1}{2} =$	$1\frac{5}{6} =$	$2\frac{1}{4} =$	$1\frac{1}{8} =$
$5\frac{1}{2} =$	$1\frac{3}{8} =$	$5\frac{1}{3} =$	$3\frac{1}{4} =$	$4\frac{1}{2} =$
$1\frac{7}{8} =$	$2\frac{2}{3} =$	$1\frac{5}{8} =$	$3\frac{3}{4} =$	$7\frac{1}{2} =$

Mental Math

a.	**b.**	**c.**	**d.**
e.	**f.**	**g.**	**h.**

Problem Solving

Understand
What information am I given?
What am I asked to find or do?

Plan
How can I use the information I am given?
Which strategy should I try?

Solve
Did I follow the plan?
Did I show my work?
Did I write the answer?

Check
Did I use the correct information?
Did I do what was asked?
Is my answer reasonable?

Name ____________________ Time __________

Power Up K

Use with Lesson 99

Facts

Complete each equivalent measure.		Write a unit for each reference.
		Metric Units:
1. 1 cm = ________ mm	13. 10 cm = ________ mm	25. The thickness of a dime: ________
2. 1 m = ________ mm	14. 2 m = ________ cm	
3. 1 m = ________ cm	15. 5 km = ________ m	26. The width of a little finger: ________
4. 1 km = ________ m	16. 2.5 cm = ________ mm	
5. 1 in. = ________ cm	17. 1.5 m = ________ cm	27. The length of one big step: ________
6. 1 mi ≈ ________ m	18. 7.5 km = ________ m	
		U.S. Customary Units:
7. 1 ft = ________ in.	19. $\frac{1}{2}$ ft = ________ in.	28. The width of two fingers: ________
8. 1 yd = ________ in.	20. 2 ft = ________ in.	
9. 1 yd = ________ ft	21. 3 ft = ________ in.	29. The length of a man's shoe: ________
10. 1 mi = ________ ft	22. 2 yd = ________ ft	
11. 1 m ≈ ________ in.	23. 10 yd = ________ ft	30. The length of one big step: ________
12. 1 km ≈ ________ mi	24. 100 yd = ________ ft	

Mental Math

a.	b.	c.	d.
e.	f.	g.	h.

Problem Solving

Understand
What information am I given?
What am I asked to find or do?

Plan
How can I use the information I am given?
Which strategy should I try?

Solve
Did I follow the plan?
Did I show my work?
Did I write the answer?

Check
Did I use the correct information?
Did I do what was asked?
Is my answer reasonable?

Name ____________________ Time __________

Power Up I

Use with Lesson 100

Facts Write each improper fraction as a mixed number. Reduce fractions.

$\frac{5}{4} =$	$\frac{6}{4} =$	$\frac{15}{10} =$	$\frac{8}{3} =$	$\frac{15}{12} =$
$\frac{12}{8} =$	$\frac{10}{8} =$	$\frac{3}{2} =$	$\frac{15}{6} =$	$\frac{10}{4} =$
$\frac{8}{6} =$	$\frac{25}{10} =$	$\frac{9}{6} =$	$\frac{10}{6} =$	$\frac{15}{8} =$
$\frac{12}{10} =$	$\frac{10}{3} =$	$\frac{18}{12} =$	$\frac{5}{2} =$	$\frac{4}{3} =$

Mental Math

a.	**b.**	**c.**	**d.**
e.	**f.**	**g.**	**h.**

Problem Solving

Understand
What information am I given?
What am I asked to find or do?

Plan
How can I use the information I am given?
Which strategy should I try?

Solve
Did I follow the plan?
Did I show my work?
Did I write the answer?

Check
Did I use the correct information?
Did I do what was asked?
Is my answer reasonable?

Name ______________________ Time __________

Power Up H

Use with Lesson 101

Facts Multiply or divide as indicated.

4×9	$4\overline{)16}$	6×8	$3\overline{)12}$	5×7	$4\overline{)32}$	3×9	$9\overline{)81}$	6×2	$8\overline{)64}$
9×7	$8\overline{)40}$	2×4	$6\overline{)42}$	5×5	$7\overline{)14}$	7×7	$8\overline{)8}$	3×3	$6\overline{)0}$
7×3	$2\overline{)10}$	10×10	$3\overline{)24}$	4×5	$9\overline{)54}$	9×1	$3\overline{)6}$	7×4	$7\overline{)56}$
6×6	$2\overline{)18}$	3×5	$5\overline{)30}$	2×2	$6\overline{)18}$	9×5	$6\overline{)24}$	2×8	$9\overline{)72}$

Mental Math

a.	b.	c.	d.
e.	f.	g.	h.

Problem Solving

Understand

What information am I given?

What am I asked to find or do?

Plan

How can I use the information I am given?

Which strategy should I try?

Solve

Did I follow the plan?

Did I show my work?

Did I write the answer?

Check

Did I use the correct information?

Did I do what was asked?

Is my answer reasonable?

Name ____________________ Time __________

Power Up M

Use with Lesson 102

Facts Write each percent as a reduced fraction and decimal number.

Percent	Fraction	Decimal	Percent	Fraction	Decimal
5%			10%		
20%			30%		
25%			50%		
1%			$12\frac{1}{2}\%$		
90%			$33\frac{1}{3}\%$		
75%			$66\frac{2}{3}\%$		

Mental Math

a.	b.	c.	d.
e.	f.	g.	h.

Problem Solving

Understand
What information am I given?
What am I asked to find or do?

Plan
How can I use the information I am given?
Which strategy should I try?

Solve
Did I follow the plan?
Did I show my work?
Did I write the answer?

Check
Did I use the correct information?
Did I do what was asked?
Is my answer reasonable?

Name ______________________ Time ____________

Power Up M

Use with Lesson 103

Facts Write each percent as a reduced fraction and decimal number.

Percent	Fraction	Decimal	Percent	Fraction	Decimal
5%			10%		
20%			30%		
25%			50%		
1%			$12\frac{1}{2}\%$		
90%			$33\frac{1}{3}\%$		
75%			$66\frac{2}{3}\%$		

Mental Math

a.	b.	c.	d.
e.	f.	g.	h.

Problem Solving

Understand

What information am I given?

What am I asked to find or do?

Plan

How can I use the information I am given?

Which strategy should I try?

Solve

Did I follow the plan?

Did I show my work?

Did I write the answer?

Check

Did I use the correct information?

Did I do what was asked?

Is my answer reasonable?

Name ______________________ Time ____________

Power Up M

Use with Lesson 105

Facts

Write each percent as a reduced fraction and decimal number.

Percent	Fraction	Decimal	Percent	Fraction	Decimal
5%			10%		
20%			30%		
25%			50%		
1%			$12\frac{1}{2}\%$		
90%			$33\frac{1}{3}\%$		
75%			$66\frac{2}{3}\%$		

Mental Math

a.	**b.**	**c.**	**d.**
e.	**f.**	**g.**	**h.**

Problem Solving

Understand
What information am I given?
What am I asked to find or do?

Plan
How can I use the information I am given?
Which strategy should I try?

Solve
Did I follow the plan?
Did I show my work?
Did I write the answer?

Check
Did I use the correct information?
Did I do what was asked?
Is my answer reasonable?

Name ____________________ Time __________

Power Up K

Use with Lesson 106

Facts

Complete each equivalent measure.		Write a unit for each reference.
		Metric Units:
1. 1 cm = ______ mm	13. 10 cm = ______ mm	25. The thickness of a dime: ______
2. 1 m = ______ mm	14. 2 m = ______ cm	
3. 1 m = ______ cm	15. 5 km = ______ m	26. The width of a little finger: ______
4. 1 km = ______ m	16. 2.5 cm = ______ mm	
5. 1 in. = ______ cm	17. 1.5 m = ______ cm	27. The length of one big step: ______
6. 1 mi ≈ ______ m	18. 7.5 km = ______ m	
		U.S. Customary Units:
7. 1 ft = ______ in.	19. $\frac{1}{2}$ ft = ______ in.	28. The width of two fingers: ______
8. 1 yd = ______ in.	20. 2 ft = ______ in.	
9. 1 yd = ______ ft	21. 3 ft = ______ in.	29. The length of a man's shoe: ______
10. 1 mi = ______ ft	22. 2 yd = ______ ft	
11. 1 m ≈ ______ in.	23. 10 yd = ______ ft	30. The length of one big step: ______
12. 1 km ≈ ______ mi	24. 100 yd = ______ ft	

Mental Math

a.	b.	c.	d.
e.	f.	g.	h.

Problem Solving

Understand
What information am I given?
What am I asked to find or do?

Plan
How can I use the information I am given?
Which strategy should I try?

Solve
Did I follow the plan?
Did I show my work?
Did I write the answer?

Check
Did I use the correct information?
Did I do what was asked?
Is my answer reasonable?

Name ______________________ Time ____________

Power Up M

Use with Lesson 107

Facts Write each percent as a reduced fraction and decimal number.

Percent	Fraction	Decimal	Percent	Fraction	Decimal
5%			10%		
20%			30%		
25%			50%		
1%			$12\frac{1}{2}\%$		
90%			$33\frac{1}{3}\%$		
75%			$66\frac{2}{3}\%$		

Mental Math

a.	b.	c.	d.
e.	f.	g.	h.

Problem Solving

Understand
What information am I given?
What am I asked to find or do?

Plan
How can I use the information I am given?
Which strategy should I try?

Solve
Did I follow the plan?
Did I show my work?
Did I write the answer?

Check
Did I use the correct information?
Did I do what was asked?
Is my answer reasonable?

Name ______________________ Time __________

Power Up M

Use with Lesson 108

Facts Write each percent as a reduced fraction and decimal number.

Percent	Fraction	Decimal
5%		
20%		
25%		
1%		
90%		
75%		

Percent	Fraction	Decimal
10%		
30%		
50%		
$12\frac{1}{2}\%$		
$33\frac{1}{3}\%$		
$66\frac{2}{3}\%$		

Mental Math

a.	b.	c.	d.
e.	f.	g.	h.

Problem Solving

Understand

What information am I given?

What am I asked to find or do?

Plan

How can I use the information I am given?

Which strategy should I try?

Solve

Did I follow the plan?

Did I show my work?

Did I write the answer?

Check

Did I use the correct information?

Did I do what was asked?

Is my answer reasonable?

Name ________________ Time ________

Power Up N

Use with Lesson 109

Facts

Complete each equivalence.

1. Draw a segment about 1 cm long.
2. Draw a segment about 1 inch long.
3. One inch is how many centimeters? ________
4. Which is longer, 1 km or 1 mi? ________
5. Which is longer, 1 km or $\frac{1}{2}$ mi? ________
6. How many ounces are in a pound? ________
7. How many pounds are in a ton? ________
8. A dollar bill has a mass of about one ________.
9. A pair of shoes has a mass of about one ________.
10. On Earth a kilogram mass weighs about ______ pounds.
11. A metric ton is ________ kilograms.
12. On Earth a metric ton weighs about ________ pounds.
13. The Earth rotates on its axis once in a ________.
14. The Earth revolves around the Sun once in a ________.

15. Water boils ______°F
16. ______°C
17. Normal body temperature ______°F
18. ______°C
19. Cool room temperature ______°F
20. ______°C
21. Water freezes ______°F
22. ______°C

°F: 40, 60, 80, 100, 120, 140, 160, 180, 200
°C: 10, 20, 30, 40, 50, 60, 70, 80, 90, 100

Mental Math

a.	b.	c.	d.
e.	f.	g.	h.

Problem Solving

Understand
What information am I given?
What am I asked to find or do?

Plan
How can I use the information I am given?
Which strategy should I try?

Solve
Did I follow the plan?
Did I show my work?
Did I write the answer?

Check
Did I use the correct information?
Did I do what was asked?
Is my answer reasonable?

Name ______________________ Time __________

Power Up M

Use with Lesson 110

Facts Write each percent as a reduced fraction and decimal number.

Percent	Fraction	Decimal	Percent	Fraction	Decimal
5%			10%		
20%			30%		
25%			50%		
1%			$12\frac{1}{2}\%$		
90%			$33\frac{1}{3}\%$		
75%			$66\frac{2}{3}\%$		

Mental Math

a.	b.	c.	d.
e.	f.	g.	h.

Problem Solving

Understand
What information am I given?
What am I asked to find or do?

Plan
How can I use the information I am given?
Which strategy should I try?

Solve
Did I follow the plan?
Did I show my work?
Did I write the answer?

Check
Did I use the correct information?
Did I do what was asked?
Is my answer reasonable?

Name ______________________ Time __________

Power Up N

Use with Lesson 111

Facts

Complete each equivalence.

1. Draw a segment about 1 cm long.
2. Draw a segment about 1 inch long.
3. One inch is how many centimeters? ________
4. Which is longer, 1 km or 1 mi? ________
5. Which is longer, 1 km or $\frac{1}{2}$ mi? ________
6. How many ounces are in a pound? ________
7. How many pounds are in a ton? ________
8. A dollar bill has a mass of about one ________.
9. A pair of shoes has a mass of about one ________.
10. On Earth a kilogram mass weighs about ______ pounds.
11. A metric ton is ________ kilograms.
12. On Earth a metric ton weighs about ______ pounds.
13. The Earth rotates on its axis once in a ________.
14. The Earth revolves around the Sun once in a ________.

15. Water boils ______°F
16. ______°C
17. Normal body temperature ______°F
18. ______°C
19. Cool room temperature ______°F
20. ______°C
21. Water freezes ______°F
22. ______°C

°F: 200, 180, 160, 140, 120, 100, 80, 60, 40
°C: 100, 90, 80, 70, 60, 50, 40, 30, 20, 10

Mental Math

a.	b.	c.	d.
e.	f.	g.	h.

Problem Solving

Understand
What information am I given?
What am I asked to find or do?

Plan
How can I use the information I am given?
Which strategy should I try?

Solve
Did I follow the plan?
Did I show my work?
Did I write the answer?

Check
Did I use the correct information?
Did I do what was asked?
Is my answer reasonable?

Name ______________________ Time ____________

Power Up M

Use with Lesson 112

Facts Write each percent as a reduced fraction and decimal number.

Percent	Fraction	Decimal	Percent	Fraction	Decimal
5%			10%		
20%			30%		
25%			50%		
1%			$12\frac{1}{2}\%$		
90%			$33\frac{1}{3}\%$		
75%			$66\frac{2}{3}\%$		

Mental Math

a.	b.	c.	d.
e.	f.	g.	h.

Problem Solving

Understand
What information am I given?
What am I asked to find or do?

Plan
How can I use the information I am given?
Which strategy should I try?

Solve
Did I follow the plan?
Did I show my work?
Did I write the answer?

Check
Did I use the correct information?
Did I do what was asked?
Is my answer reasonable?

Name ______________________ Time __________

Power Up N

Use with Lesson 113

Facts

Complete each equivalence.

1. Draw a segment about 1 cm long.
2. Draw a segment about 1 inch long.
3. One inch is how many centimeters? ______
4. Which is longer, 1 km or 1 mi? ______
5. Which is longer, 1 km or $\frac{1}{2}$ mi? ______
6. How many ounces are in a pound? ______
7. How many pounds are in a ton? ______
8. A dollar bill has a mass of about one ______.
9. A pair of shoes has a mass of about one ______.
10. On Earth a kilogram mass weighs about ______ pounds.
11. A metric ton is ______ kilograms.
12. On Earth a metric ton weighs about ______ pounds.
13. The Earth rotates on its axis once in a ______.
14. The Earth revolves around the Sun once in a ______.

15. Water boils ______°F
16. ______°C
17. Normal body temperature ______°F
18. ______°C
19. Cool room temperature ______°F
20. ______°C
21. Water freezes ______°F
22. ______°C

°F: 40, 60, 80, 100, 120, 140, 160, 180, 200
°C: 10, 20, 30, 40, 50, 60, 70, 80, 90, 100

Mental Math

a.	b.	c.	d.
e.	f.	g.	h.

Problem Solving

Understand
What information am I given?
What am I asked to find or do?

Plan
How can I use the information I am given?
Which strategy should I try?

Solve
Did I follow the plan?
Did I show my work?
Did I write the answer?

Check
Did I use the correct information?
Did I do what was asked?
Is my answer reasonable?

Name ______________________ Time ____________

Power Up M

Use with Lesson 114

Facts

Write each percent as a reduced fraction and decimal number.

Percent	Fraction	Decimal	Percent	Fraction	Decimal
5%			10%		
20%			30%		
25%			50%		
1%			$12\frac{1}{2}\%$		
90%			$33\frac{1}{3}\%$		
75%			$66\frac{2}{3}\%$		

Mental Math

a.	b.	c.	d.
e.	f.	g.	h.

Problem Solving

Understand

What information am I given?

What am I asked to find or do?

Plan

How can I use the information I am given?

Which strategy should I try?

Solve

Did I follow the plan?

Did I show my work?

Did I write the answer?

Check

Did I use the correct information?

Did I do what was asked?

Is my answer reasonable?

Name ______________ Time ______

Power Up N

Use with Lesson 115

Facts

Complete each equivalence.

1. Draw a segment about 1 cm long.
2. Draw a segment about 1 inch long.
3. One inch is how many centimeters? ______
4. Which is longer, 1 km or 1 mi? ______
5. Which is longer, 1 km or $\frac{1}{2}$ mi? ______
6. How many ounces are in a pound? ______
7. How many pounds are in a ton? ______
8. A dollar bill has a mass of about one ______.
9. A pair of shoes has a mass of about one ______.
10. On Earth a kilogram mass weighs about ______ pounds.
11. A metric ton is ______ kilograms.
12. On Earth a metric ton weighs about ______ pounds.
13. The Earth rotates on its axis once in a ______.
14. The Earth revolves around the Sun once in a ______.

15. Water boils ______ °F
16. ______ °C
17. Normal body temperature ______ °F
18. ______ °C
19. Cool room temperature ______ °F
20. ______ °C
21. Water freezes ______ °F
22. ______ °C

°F: 200, 180, 160, 140, 120, 100, 80, 60, 40
°C: 100, 90, 80, 70, 60, 50, 40, 30, 20, 10

Mental Math

a.	b.	c.	d.
e.	f.	g.	h.

Problem Solving

Understand
What information am I given?
What am I asked to find or do?

Plan
How can I use the information I am given?
Which strategy should I try?

Solve
Did I follow the plan?
Did I show my work?
Did I write the answer?

Check
Did I use the correct information?
Did I do what was asked?
Is my answer reasonable?

Name ______________________ Time __________

Power Up M

Use with Lesson 116

Facts Write each percent as a reduced fraction and decimal number.

Percent	Fraction	Decimal
5%		
20%		
25%		
1%		
90%		
75%		

Percent	Fraction	Decimal
10%		
30%		
50%		
$12\frac{1}{2}\%$		
$33\frac{1}{3}\%$		
$66\frac{2}{3}\%$		

Mental Math

a.	b.	c.	d.
e.	f.	g.	h.

Problem Solving

Understand

What information am I given?

What am I asked to find or do?

Plan

How can I use the information I am given?

Which strategy should I try?

Solve

Did I follow the plan?

Did I show my work?

Did I write the answer?

Check

Did I use the correct information?

Did I do what was asked?

Is my answer reasonable?

Name ______________________ Time __________

Power Up N

Use with Lesson 117

Facts

Complete each equivalence.

1. Draw a segment about 1 cm long.
2. Draw a segment about 1 inch long.
3. One inch is how many centimeters? ______
4. Which is longer, 1 km or 1 mi? ______
5. Which is longer, 1 km or $\frac{1}{2}$ mi? ______
6. How many ounces are in a pound? ______
7. How many pounds are in a ton? ______
8. A dollar bill has a mass of about one ______.
9. A pair of shoes has a mass of about one ______.
10. On Earth a kilogram mass weighs about ______ pounds.
11. A metric ton is ______ kilograms.
12. On Earth a metric ton weighs about ______ pounds.
13. The Earth rotates on its axis once in a ______.
14. The Earth revolves around the Sun once in a ______.

15. Water boils ______°F
16. ______°C
17. Normal body temperature ______°F
18. ______°C
19. Cool room temperature ______°F
20. ______°C
21. Water freezes ______°F
22. ______°C

°F: 200, 180, 160, 140, 120, 100, 80, 60, 40
°C: 100, 90, 80, 70, 60, 50, 40, 30, 20, 10

Mental Math

a.	b.	c.	d.
e.	f.	g.	h.

Problem Solving

Understand
What information am I given?
What am I asked to find or do?

Plan
How can I use the information I am given?
Which strategy should I try?

Solve
Did I follow the plan?
Did I show my work?
Did I write the answer?

Check
Did I use the correct information?
Did I do what was asked?
Is my answer reasonable?

Name ______________________ Time ____________

Power Up M

Use with Lesson 118

Facts Write each percent as a reduced fraction and decimal number.

Percent	Fraction	Decimal	Percent	Fraction	Decimal
5%			10%		
20%			30%		
25%			50%		
1%			$12\frac{1}{2}$%		
90%			$33\frac{1}{3}$%		
75%			$66\frac{2}{3}$%		

Mental Math

a.	b.	c.	d.
e.	f.	g.	h.

Problem Solving

Understand
What information am I given?
What am I asked to find or do?

Plan
How can I use the information I am given?
Which strategy should I try?

Solve
Did I follow the plan?
Did I show my work?
Did I write the answer?

Check
Did I use the correct information?
Did I do what was asked?
Is my answer reasonable?

Name ______________ Time ________

Power Up N

Use with Lesson 119

Facts

Complete each equivalence.

1. Draw a segment about 1 cm long.
2. Draw a segment about 1 inch long.
3. One inch is how many centimeters? ________
4. Which is longer, 1 km or 1 mi? ________
5. Which is longer, 1 km or $\frac{1}{2}$ mi? ________
6. How many ounces are in a pound? ________
7. How many pounds are in a ton? ________
8. A dollar bill has a mass of about one ________.
9. A pair of shoes has a mass of about one ________.
10. On Earth a kilogram mass weighs about ________ pounds.
11. A metric ton is ________ kilograms.
12. On Earth a metric ton weighs about ________ pounds.
13. The Earth rotates on its axis once in a ________.
14. The Earth revolves around the Sun once in a ________.

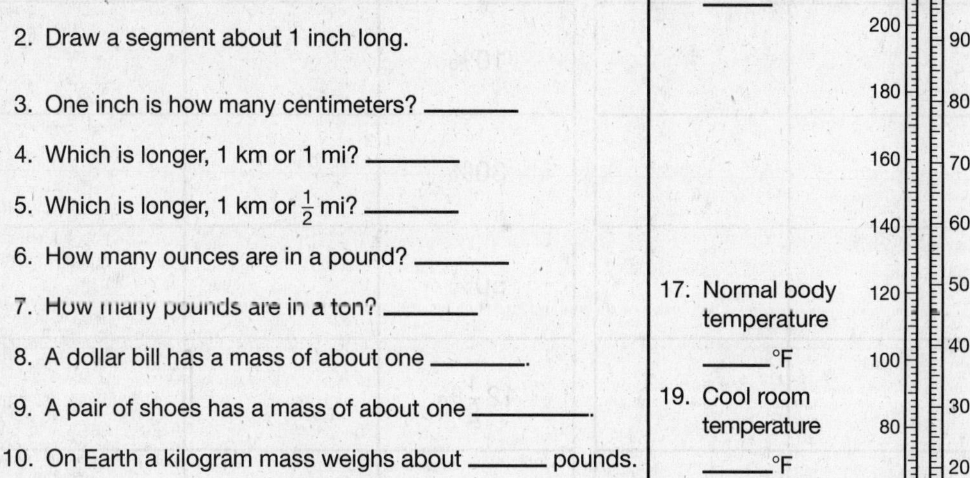

15. Water boils ______°F
16. ______°C
17. Normal body temperature ______°F
18. ______°C
19. Cool room temperature ______°F
20. ______°C
21. Water freezes ______°F
22. ______°C

Mental Math

a.	b.	c.	d.
e.	f.	g.	h.

Problem Solving

Understand

What information am I given?
What am I asked to find or do?

Plan

How can I use the information I am given?
Which strategy should I try?

Solve

Did I follow the plan?
Did I show my work?
Did I write the answer?

Check

Did I use the correct information?
Did I do what was asked?
Is my answer reasonable?

Name ______________________ Time __________

Power Up M

Use with Lesson 120

Facts Write each percent as a reduced fraction and decimal number.

Percent	Fraction	Decimal	Percent	Fraction	Decimal
5%			10%		
20%			30%		
25%			50%		
1%			$12\frac{1}{2}\%$		
90%			$33\frac{1}{3}\%$		
75%			$66\frac{2}{3}\%$		

Mental Math

a.	b.	c.	d.
e.	f.	g.	h.

Problem Solving

Understand

What information am I given?

What am I asked to find or do?

Plan

How can I use the information I am given?

Which strategy should I try?

Solve

Did I follow the plan?

Did I show my work?

Did I write the answer?

Check

Did I use the correct information?

Did I do what was asked?

Is my answer reasonable?